FETAL
BEAUTY

FINDING JUSTICE IN THE WOMB

Jordan Lynn Warfel

Illustrations by Marco Chiu
Cover design by PixelStudio
Layout design by Inkcept Studio
Editing by Natalie Groff

Cover Photography by Stephen O'Connor, MD
August 7, 2013 Houston, TX
Used with permission
All rights reserved
6-7 week spontaneous loss

This book includes detailed descriptions of abortion procedures in chapters 11-16. Reader discretion is advised. This book does not contain pictures of aborted children. Nothing in this book should be construed as medical or legal advice. If you are struggling with the loss of a child, please seek professional help.

First paperback edition: December 2019
ISBN: 978-1-7344108-1-5
Candlelight Publishing
Greenwood, Delaware

Jordan Lynn Warfel

FETAL BEAUTY

FINDING JUSTICE IN THE WOMB

Candlelight Publishing
Greenwood

ABOUT THE AUTHOR

Jordan Warfel resides in Delaware with his wife Felicia and their three children. Jordan was born and raised in southern Delaware where his parents were board members for the local pro-life pregnancy care center. After graduation, he achieved an associate of arts degree in Biblical studies from Rosedale Bible College and a bachelor of science degree in organizational management from Wilmington University. Since 2005 Jordan has been involved with numerous aspects of the pro-life movement including legislation, political campaigns, and media. Today Jordan is the president of the annual Candlelight Walk for Life in Milford, Delaware and regularly gives pro-life presentations to churches and community groups. When he is not working as a residential draftsman and in the pro-life movement, he enjoys playing with his kids and making improvements to their home.

SPEAKING REQUESTS

Jordan Warfel is available to speak to your church, school, or community group. You can request a hands-on interactive presentation for audiences of all sizes based on the content in *Fetal Beauty*. Presentations are affordable and can be tailored to the interests of your audience. To book a presentation, send an email to jordanwarfel@gmail.com.

www.FetalBeauty.com

DEDICATION

This book is dedicated to Nicole,
for entrusting us to adopt your son.
Our society is brimming with messages telling
you that abortion is the solution.
But you had the courage to do
what was best for your unborn child.
We owe a debt to you that we can never repay.
When we think of heroes,
we think of you.

CONTENTS

INTRODUCTION

An unusual meeting is taking place at a small Quaker meeting house in Great Britain in the late 18th century. A tall red haired man by the name of Thomas Clarkson is standing and giving a lecture. He is a well-educated intellectual known for his award winning Latin essay in opposition to the slave trade. In front of the crowd is a specially made wooden box designed to hold Clarkson's evidence. One by one Clarkson removes items from his wooden box. He shows his audience the valuable and ingenious products from Africa: products he claims Britain is not able to enjoy due to the slave trade. Then he lifts out the tools of the trade. The lecture turns from curiosity to horror as he demonstrates how the tools of the trade, tools like the thumb screw and shackles, are used on African bodies.

Clarkson gave countless lectures like this throughout Great Britain. This man was one of the most recognizable abolitionists in Great Britain. Over his lifetime, he traveled tens of thousands of miles, often putting his own safety and health at risk, to collect evidence and testimonies about the Atlantic slave trade. Most of his life was dedicated to this work.

The reason his work was so vitally important was because he took the mask off of slavery. It's one thing to have a debate about the merits of the slave trade in the context of politics. It's another thing entirely to see the nature of the violence for yourself. While slavery was hotly debated in Great Britain, the public had to be educated about the true nature of the violence perpetrated against Africans. Slavery's mask had to be removed. Clarkson was successful in revealing to the public what slavery truly was. It was no longer a theoretical political debate. It was real human bodies being subjected to unspeakable violence. Thanks to Clarkson's work, the abolition movement went from being limited to small religious minorities like the Quakers to a broadly popular movement. If you'd like to learn more about Thomas Clarkson, I highly recommend *Biographical Sketch of Thomas Clarkson* written by Thomas Taylor in 1839.[1]

After reading about Clarkson, it occurred to me that America has much the same problem with abortion today as the British had with slavery then. Conversations around abortion are largely limited to political rhetoric and familiar talking points. There is little understanding of the unborn children and the procedures performed on them. We shouldn't be surprised then that pollster George Barna called America's views on abortion "confused and lukewarm." According to Barna, most Americans hold views on abortion that are internally inconsistent. For example, he found that 23% of self-described pro-lifers believe that abortion should be legal in all or most cases. He further found that 25% of self-described pro-choice people also believe that abortion at any stage is murder.[2]

The parallels between Thomas Clarkson's culture and our culture are striking. It's not just that the pro-life movement tends to be limited to Evangelicals and devout Catholics in the same way that abolition tended to be limited to the Quakers. Most Americans debate abortion in theory, but have a limited understanding of its violent nature. Clarkson gave me the inspiration to understand abortion better for myself. I bought the textbooks used to train abortionists. I bought fetal models and the tools of the abortion trade. Then I began doing my own demonstrations. In my abortion demonstrations participants are able to see abortion for what it truly is. Clarkson demonstrated how the thumb screws and shackles were used on the slave bodies. I demonstrate how the curettes and forceps are used on the tiny unborn bodies. Clarkson humanized the Africans by showing their ingenuity and the products they produce. I humanize the unborn children by showing their highly-sophisticated bodies and their wondrous development from the moment of conception.

Abortion is wearing a mask. The mask's purpose is to give abortion the appearance of compassion and empathy for women. But the mask isn't reality. The mask is covering up the reality. It's time that we remove the mask and look abortion violence in the eye. At the heart of the abortion debate is the question "Do the difficult and tragic situations that women with unwanted pregnancies face justify a violent abortive response?" It's only after we stare down the violence and reject it that we can become a truly equal and just society.

Let me say at the outset that this book is not written to be shocking or gory. My goal is to leave you the reader better educated to make your own decisions about abortion based on the facts, the science, and the medicine. Sometimes the facts are hard to hear. But a well-educated opinion must consider what an abortion actually is. This is not a book for pro-life people. This is not a book for pro-choice people. This is a book for everyone who loves equality and justice under the law. This is a book for anyone who loves peace and rejects violence. This book is a call to all Americans of good faith to live up to our core values: peace, justice, and equality.

Introduction Notes

1. Taylor, Thomas. A *biographical sketch of Thomas Clarkson, M.A. with occasional brief strictures on the misrepresentations of him contained in the Life of William Wilberforce [by Wilberforce's sons]; and a concise historical outline of the abolition of slavery.* London: J. Rickerby. (1839)

    This book can be read for free online at https://catalog.hathitrust.org/Record/001741958

2. Barna, George. "Americans Confused About Abortion." (2017)

    http://www.georgebarna.com/research-flow/2017/9/27/americans-confused-about-abortion

Part A

OUR FAMILY'S STORY

Fetal Beauty

NEVER DESPAIR

Chapter 1

Our Foster Care Journey

---

*Never despair of a child. The one you weep
the most for at the mercy seat may fill your
heart with the sweetest joys.*[1]

Theodore L Cuyler

Years before Felicia and I started dating, Felicia sent me a satirical email telling me of her "life schedule." The supposed schedule included thirteen kids. The whole thing was just a playful joke between friends. But when I decided that I wanted to be more than friends, I hit reply to that old email and changed the course of our lives. When Felicia read my reply, she only read the first paragraph in which I told her that she was in the part of her schedule in which she needs to find a husband. She skipped the second half in which I asked to date her. She replied back continuing the silly joke, not knowing I actually wanted to date! We still laugh about that awkward beginning. But our marriage has been anything but a joke.

When we reached that blissful moment at the altar, there were so many thoughtful plans for our future and how we would grow our family. For most people in our country it goes something like this: you get married, you settle down in a nice community, you have two or three babies, and you go on to live the "American dream." All too often those well thought out plans are interrupted by years of bloodwork, embarrassing tests, injections, and negative pregnancy tests. This is the harsh reality of infertility: a common struggle that affects one in every eight couples trying to get pregnant and carry the baby successfully to birth.[2] We didn't plan to have our marriage so humiliatingly interrupted. But unfortunately, that is exactly what happened to us.

Our story of infertility spans several difficult years. Interspersed with that story are our years as foster parents and then adoptive parents. Allow me to share our stories of foster care in chapter one and adoption in chapter two. In chapter three, Felicia will share about our journey through fertility treatments which eventually gave us two biological daughters.

Not being ones to shy away from hard work, we struggled through infertility while also signing up to be foster parents. Felicia and I both had an interest in foster care and adoption before we married. Starting the journey, therefore, was an easy decision. Foster care is a maddening process. It is your job as a foster parent to advocate for your child's best interest, not just to parent them while they are placed with you. We took the role of advocates very seriously. I made emotional appeals at court hearings and committee meetings. In my experience and in the experience of many of my friends in the Delaware foster care system, mistakes are common and not everyone in the system is looking out for the best interest of the child. More often than not, it felt like we were fighting a system that was stacked against our kids.

Our last placement of a baby girl was a prime example. The mistakes made in her case were infuriating. At one point her social worker claimed in court that she had filed for termination of parental rights. We found out months later in court that she never did file. This kind of substantial and willful lie under oath is perjury. As is often the case, the worker was never held accountable. The mistake resulted in our foster daughter having three different goals in court: termination of parental rights, reunification with the biological mother, and guardianship with a relative. If the court had pursued adoption, we would have gladly adopted her. Because of the legal confusion, precious time was wasted and it gave state workers the opportunity to pursue a goal that was not in her best interest. It took a new judge months later to straighten out the mess our foster care system had created.

Many times as a foster parent I found myself needing to take on the role of a social worker. In order to pursue one child's best interest I had to contact her relatives in another state who wanted guardianship, and help them where the state had failed them. I had to notify them of court hearings because the workers failed to do so. I had to educate them on our legal system and how to file for guardianship. I even had to help fill out the guardianship papers for them and arrange to have a social worker meet them at the family court to file for guardianship. My job as a foster parent felt like that of a parent, social worker, and attorney all rolled into one. It was hard work, but I loved it when it wasn't tearing me apart.

As foster parents, we took placements and provided respite care. A placement is when the foster child is placed in your home, often until the case is finally resolved. Respite care is when you take a child overnight for as long as one to two weeks to give the foster parents a break. Our first placement wasn't technically considered a placement. We took them for respite and the girls ended up staying with us for two months. We also provided babysitting. If you count them, we had more than a dozen children in our care. In nearly four years we cared for two sibling sets of three, twins with a genetic disability, and placement of a little girl that lasted over a year. We even took an emergency placement. They called us at 3:00 in the afternoon. I agreed to take placement without even asking Felicia. A few short hours later we had an exhausted one year old falling asleep in the middle of our living room floor. We also got calls for placements we couldn't take. On one occasion we were called for an emergency placement of a seventeen year old girl. We had to say no to that one. When you are a foster parent, you quickly learn that you can't say yes to every phone call. But it's always an adventure.

Foster care is inherently difficult. For couples like us who couldn't conceive, it is even more difficult. You can't help but hope that your foster child will be legally available for adoption. We tried our best not to get our hopes up but it was futile. We ultimately never had the chance to adopt any of our foster children.

The social workers know whether or not you are open to adoption. Many times they play the adoption card when they are having a hard time placing children. On one occasion we agreed to take placement of a sibling set of three. The worker claimed that they would be filing for termination of parental rights within days and we would have the opportunity to adopt them. It wasn't remotely true. They had a court hearing a few days later where the worker pursued reunification with the biological parents. They were not pursuing termination of parental rights. I was fortunate enough to meet their aunt a couple of years later and learn that they had been successfully reunited with their biological Father. There is no way to see into the future of these foster children. You can only lean on God and pray that He works it all out for good.

On another occasion, we agreed to take placement of twin girls with a genetic disability. Having children with disabilities brings extra challenges. These toddlers were about thirty pounds each and couldn't walk. We had to do special exercises each day to try to teach them to walk. The girls hated their exercises. Further, their nursery was upstairs which meant carrying them up and down the stairs every time they slept. One of the things I learned from these girls was how special children with disabilities are and how important it is that we give them an opportunity to have a meaningful life. We loved our time with the twins. It breaks my heart that so many of these disabled little ones are unwanted and discarded through abortion.

With the twins, a worker again played the adoption card on us. There was a lot of pressure to find an adoptive family for them. We made it clear when we took them in that we were not qualified to adopt them. Despite making our intentions clear, we still received a call from a manager at our private agency trying to pressure us into adopting them. We stuck to our decision that we couldn't adopt them. By the end of the conversation she dropped the threat. "If you don't adopt them, I'm not sure I can place more children with you." I've had some people question what she really meant. But within the context of our conversation, I think she was quite clear. It was a threat.

I've had many conversations with a variety of people about the effectiveness of Delaware's foster care system. These people ranged from a policy advisor for a Delaware cabinet secretary to an applicant for a position in Delaware Family Services. Unfortunately, I can't say that our foster care system is particularly effective. The standard under Delaware law is the best interest of the child. Are we always acting in the best interest of every child in state custody? I expect nothing less in every single case. Unfortunately, in my experience, those representing the state often do not act in the child's best interest. It is common for children to be kept in foster care for longer than is necessary. Government social workers pursue goals they shouldn't pursue. These workers aren't always honest with foster parents and even apparently with the judge. The list goes on and on, each with real world consequences for these kids.

So what is the cause and the solution? Unfortunately, the solutions we've tried seem to be making the system worse not better. The response to the problems has been to create more bureaucracy. We have committees on top of committees. And many of these committees don't seem to be taken seriously. Is all this bureaucracy helping or making things worse? I tend to think it is just creating more busy work. What's really needed in my humble opinion is real accountability. I even recommended to Governor Carney's administration that workers involved in placing children be licensed by the state and be required to follow a code of ethics similar to many other professions. If a worker lies to a foster parent, that worker should not be allowed to place children. The same goes for workers who disregard a child's goal. If you aren't willing to pursue a child's legal goal, or seek to change that goal when appropriate, you shouldn't be a social worker. Until we have real consequences for individuals who aren't doing what's in the best interest of our foster children, you can expect our system to continue to fail them.

Despite the failings of the system, many good people throughout are working hard for these kids. Our time as foster parents was often frustrating and stressful. But we miss it still and look forward to being able to foster again. In the three and a half years that we provided foster care, we had a dozen children through our home. I treasured every moment. I miss the laughter, the hugs, and the smiles. I miss being their Daddy and protector. I have a lot of people tell me that they don't think they could be foster parents because it means becoming emotionally attached to children who will have to leave. What I wished that they all knew is that the good times more than make up for the hard times. Becoming emotionally attached is not only necessary for the healthy development of your foster children, but is also rewarding. The attachment you build with the kids more than makes up for the depression you experience when they leave. We have a wall of pictures of those little ones in our kitchen. Every so often I look at the pictures and remember those good days. I don't have any regrets.

For anyone considering foster care, I'd encourage you to do it. Consider the cost first, especially the emotional and psychological cost. Understand what you are signing up for. Make sure your heart is in the right place. Then throw caution to the wind and sign up. It is a highly rewarding and profoundly fulfilling responsibility.

Chapter Notes

1. Robinson, Charles S, DICTIONARY OF BURNING WORDS. New York: Wilbur Ketcham, 1895. pp. 50

    Theodore L. Cuyler (1822-1909) was a popular Presbyterian preacher, author, and abolitionist.

2. The National Infertility Association. "Get The Facts" https://resolve.org/infertility-101/what-is-infertility/fast-facts/

BRINGING HOME DERRICK

Chapter 2

Our Adoption Journey

After Felicia and I had done several failed rounds of fertility treatments, and we hadn't had the opportunity to adopt through foster care, we had a defining moment of crisis. We were at a major crossroads that would determine the future of our family for generations. Fertility treatments had failed and foster care left us physically and emotionally exhausted. We had to consider all our options, including continuing foster care, an expensive domestic infant adoption, or being child free. For a brief time, we seriously explored the benefits of being child free. We talked about Felicia going back to nursing school. We talked about all the deeply meaningful things we could do without kids. But we weren't sold on the idea. We still had some fight left in us. One morning in January of 2015 we were sitting on the front row at church when Felicia leaned over and slipped me a note. On the back of the church bulletin she wrote, "I want to adopt a baby." In that moment we knew exactly what we needed to do. It was one of those rare and inexplicable moments of divine clarity.

Let me start by saying that the adoption process in our situation went unusually smoothly. I don't want to give the false impression that it is always as easy as it was for us. We pursued a private domestic infant adoption. What that means is that a private adoption agency matched us with a pregnant woman to adopt her baby. If you are considering adoption, as of the time of this writing, it is becoming increasingly difficult to do this type of adoption. There are several factors making it more difficult. One such factor is rising infertility in the United States. More infertile couples means a greater demand for newborn babies. Due to America's unusually permissive abortion laws which allow abortion for any reason until 20-24 weeks and sometimes even through all nine months of pregnancy, we have had a shortage of newborn babies to adopt going back decades. To make matters more difficult for prospective adoptive couples, increased social acceptance of single parents means that more women are choosing parenting over adoption. The result is a perfect storm resulting in far more perspective adoptive parents than available babies. The prospects for foreign adoptions aren't any better. According to Nightlight

Christian Adoptions, a private adoption agency, foreign adoptions in the United States dropped from 24,000 successful adoptions in 2008 to only 4,700 successful adoptions in 2017.[1] The precipitous decline in foreign adoptions means that the overwhelming demand for children to adopt is affecting the entire adoption community.

For us the adoption started with the home study. Because we were already with an agency for foster care, we were able to get our home study much more quickly than most perspective parents. Home studies are almost always required for adoptions in the United States. A social worker writes the home study to determine if the perspective person or couple is qualified to adopt and what kinds of children they are qualified to adopt. It is a written report that goes into great detail about your family. Home studies can require things like a physical examination by a doctor, vast amounts of documentation, and an inspection of the home. For some people the home study can take months. This is usually the first step in the adoption process before you can move forward with the adoption.

In our situation, we decided to wait to sign up with an agency to be matched with a birth mom until the home study was done. This gave us time to shop around the agencies and find out which ones had the shortest wait times to be matched. Different agencies work differently. Some agencies are very exclusive. They have their own birth moms that they work with, and only the parents with that agency can be matched with those birth moms. It is not unusual to have your profile shown to only one or two birth moms each month. Felicia and I chose an agency that had access to a large nationwide network of birth moms. We had our profile shown to a birth mom almost every day, and sometimes twice a day.

The next step for us was creating our profile book. Each agency has its own rules for profile books. Our book was about 25 pages long. It had lots of pictures and information about our family and what we had to offer a child. Good advice about profile books is hard to find. We discovered that most agencies don't tend to review the books with a critical eye. One of the best decisions we made was to buy a book from a professional profile book writer. *How to Create a Successful Adoption Portfolio* by Madeleine Meicher was great reading while we were waiting for our home study to be written. She goes in depth into the thought processes of birth moms and how to write a profile that appeals to them. She also offers to review your profile for free if you buy her book. We took her up on that offer and were blown away by how much great advice she gave us. I believe she was the only person to review our profile with a truly critical eye. The last thing I wanted was to have people tell me what they thought I wanted to hear. I wanted to know what was wrong with it so I could make it better.

After our profile books were sent to the agency, we began the waiting. Waiting can be a stressful job. There is nothing you can do at this point. We did all the usual stuff like buying cute outfits and decorating the nursery. We spent lots of time on the rocking chair in that nursery waiting for our baby. Then one day when I least expected it, we got the phone call. An expectant mother in Georgia picked our profile. Being picked is just one step in getting matched. The next step was to talk to her on the phone. I headed straight to Walmart for a burner cell phone. We called it our adoption phone. This was so that we could communicate with her without giving her our personal phone numbers. The adoption phone went with us everywhere until Derrick was born.[2] After his birth, we kept the phone active for a year so that it would be possible to continue communication with his biological mom. We call her "Tummy Mommy" and Felicia we call "Forever Mommy."

We decided with our adoption to keep some lines of communication. You may have heard of open adoptions and closed adoptions. Closed adoptions aren't nearly as common today as they once were. In a closed adoption there is no communication and no information is known about the people involved. It is common for kids in closed adoptions to know nothing about their biological families. That includes not knowing the medical history of the biological family. I personally would not recommend a closed adoption to anyone unless it is necessary for safety reasons. Open adoptions are a much broader category of adoptions. This could be as little as some communication before the birth or as much as having the birth mom involved throughout the child's life. In private domestic adoptions, the amount of communication and information shared is often negotiated between the expectant mother and prospective adoptive family. The agreement we negotiated was that we would keep the adoption phone active for one year and keep our anonymous email account open indefinitely. We agreed to send her pictures occasionally at her request. His birth mom has been very considerate. We've exchanged a few emails and pictures since the adoption. We hope to continue to hear from her occasionally so that we can share information with Derrick as he gets older.

After we were picked, the next step was to talk to her over the phone. I can't remember exactly what was said in that first phone call. It was all a blur. I just remember that she made up her mind that she wanted us and hoped that we would agree to adopt her baby. We didn't have to convince her. We were all sold on the match from the beginning. All we had to do was wait for medical records so that the match would become official.

The next step for us was to make a road trip to meet her. The emotions are so unique to this situation that I have no way to describe them. It was a weekend trip. We drove down 12 hours, met her for lunch, and drove back. We went to her favorite place, a local seafood restaurant. I was adventurous and tried the alligator. We talked about how we would raise him. I told her that I played piano. She was excited at the prospect of him being a piano player. She told us she liked Alicia Keys. The whole experience was surreal, like a dream. Was this really happening to us? Was this lady really giving us her baby? On the way back to her grandmother's house, Felicia asked if she could feel the baby kick. Tummy Mommy put Felicia's hand on her belly. I thought Felicia was going to lose it. She got to feel our baby kick!

Our next trip down was for a scheduled caesarian section. Again the whole trip was like a dream. I had to keep checking myself to see if it was real. She let us pick his name. That's a really big deal. She had a legal right to name him anything she wanted. He was her baby. And we had a right to change his name when we adopted him. There wasn't even any discussion. She just asked us what we wanted to name him. We told her

Derrick after a certain famous person that we admire. She seemed to like it and immediately started calling him her little Derrick. Now you can see why I dedicated this book to her. I saw a side of her most people likely didn't see. I saw an unusually strong woman who rose to the occasion and conducted herself with grace and dignity.

On that Monday morning we picked her up from her Grandmother's house and took her to the hospital. There was no family to support her. We were her family. Felicia and I were in the delivery room when he was born. I held Tummy Mommy's hand during the surgery. We even had our own room in the maternity ward. Until she was discharged, we would take frequent trips back and forth between our room and Tummy Mommy's room. When she was discharged, our attorney/social worker physically took Derrick from birth mom and placed him with us. We took placement of Derrick right outside the hospital's front doors. The next day we met birth mom and her other son for breakfast at IHOP while we waited for the legal paperwork that would allow us to leave the State of Georgia. Derrick was born on Monday and by Friday afternoon we were leaving the state. I'm sure some day we will meet his tummy mommy again when Derrick is ready. Today we are a chaotically adventurous family of five. It's been a long, difficult, and worthwhile journey.

There are two reasons that I share our foster care, adoption, and infertility story with you.

The first reason is to tell you that parenting is worth it. You can't know how truly great parenting is until you try it. Life is hard. You may be a single parent. You may be living on welfare. You may have very little support. But it's worth it. If you find yourself unexpectedly pregnant, and you can find any possible way to parent, just do it. I'm absolutely convinced you will be glad you did. And if parenting isn't possible, as it wasn't for Derrick's tummy mommy, then you can be a hero like her and choose adoption. You can find the strength within yourself to do what's best for your child.

The second reason I tell our story is so that you can see that I am personally acquainted with tragedy. Unfortunately, many pro-choice people live in a bubble where they falsely believe that pro-life people are pro-life because they don't understand how hard things are for women with unplanned pregnancies. They believe that if we really understood the tragedy, that we would feel enough sympathy for the mother to support her right to an abortion. I can tell you from personal experience that the opposite is true. Of course we feel sympathy for women in crisis. But we rightly understand that abortion is its own tragedy, not a solution to tragedy.

Most pro-choice people don't openly claim that violence is the solution to an unwanted pregnancy. But that is exactly what an abortion is. We are a society that has chosen an inherently violent act, which we call abortion, as the solution to unwanted pregnancies and the tragedies that go along with them. The pro-life message is simple. Violence must never be the answer. Rejecting violence and embracing peace is the heart and soul of the pro-life movement. If you reject violence and love peace, then we invite you into the pro-life community. You belong with us.

Chapter Notes

1. Nehrbrass, Daniel. "Why are international adoptions on the decline?" Nightlight Christian Adoptions (2017) https://www. nightlight.org/2017/10/international-adoptions-decline/

2. His name has been changed to protect his privacy.

Fetal Beauty

SONGS OF JOY

Chapter 3

Our Infertility Journey
Written by Felicia Warfel

---

*Those who sow with tears will reap with
songs of joy. Those who go out weeping,
carrying seed to sow, will return with
songs of joy, carrying sheaves with them.*

Psalm 126: 5-6 NIV[1]

I was nearing my 26th birthday when I said "I do" to
my best friend. I would be lying if I failed to admit
that high on the list of reasons I was excited to get
married was because it put me one step closer to being
a mother. When we eventually started trying to get
pregnant I was excited and filled with anticipation of
what was about to happen. After the first month I didn't
feel very much frustration. Even after the second and
third months I wasn't concerned. The idea that I should
be didn't occur to me. Medical doctors say that it can
take a few months to get pregnant. Eventually when a
few months turned into many, I became very frustrated.
Looking back, I can't remember exactly what I was

thinking when the doctor referred us to a fertility specialist, but I know that the feelings I experienced were ones that I would relive again and again over the next five years. No one warned me that those feelings, the uncertainty, frustration and worries, would only worsen over time.

Our first visit to the fertility specialist was overwhelming. The doctor sat with a diagram of the female reproductive system and outlined everything that can go wrong and the tests to diagnose those problems. With our minds bulging with information, we sat down with the nurse and filled up my calendar for the next month with appointments for lab work, tests, and ultrasounds. I tried to work this around my part-time work schedule and the fretting that would take over the next few years started. How would I maintain my privacy at work while having all these appointments?

A month later, with the final test results back, we consulted with the fertility doctor and made a plan. It was time to start the cycle: medications, ultrasounds, and lab work. This was repeated several days a week for the next nine months. Each month, when the timing was correct, we had the anticipated appointment for Intrauterine Insemination (IUI). Ten days later, without fail, we got the call that the pregnancy test done in the lab was negative. We would reevaluate, change medication dosages, pray more, and cry more. After several months we elected to do a diagnostic surgery which revealed silent endometriosis. It is called silent endometriosis because I had no symptoms other than infertility. But despite surgical treatment, those difficult phone calls continued.

You can only go on so long like this. I was forced to be evasive with my coworkers and acquaintances, not wanting to share this private part of my life. I made excuses and struggled with not wanting to outright lie to them. Often on mornings I had appointments, I was a few minutes late to work. Eventually I was forced to share confidentially with my boss what we were doing so that there would be less conflict.

My time as a patient of the fertility specialist came to an end after yet another consult. He told us it wasn't working and that we should move on to something else. He recommended In Vitro Fertilization (IVF). Discouraged, we left the appointment with a packet of information on the process and financial options. There was no insurance coverage available for the actual process, just the diagnostics, and it was costly.

It took a long weekend away at our favorite place in the world, Shenandoah National Park, to make the decision to walk away from fertility treatments. The cost was overwhelming. The emotional stress of continuing treatments was too hard to bear. All of those concerns aside, we had many questions about the ethics of IVF that we couldn't resolve. Our time pursuing pregnancy through fertility treatments came to an end, at least for the time being.

It was the late summer of 2014 when we made that difficult, but also relieving, decision. We took the next 6 months to recover and start a healing process. At the end of that, still desiring a child and ready to pursue another route, we began the adoption process (detailed in chapter 3). Through it all though, in the back of my mind, I knew I wanted to go back, to try a few more IUIs, a few more drugs, and a whole lot more prayers.

Let's skip ahead to when Derrick, our adopted son, was about six months old. We'd been discussing fertility treatments again and planning on waiting until after Derrick's first birthday. We were concerned that I was no longer needed at my job because of my part-time status. Knowing that I might lose my health insurance and what little assistance it gave, we made an appointment with the fertility specialist. That appointment was very similar to our first: the same drawing of the reproductive system, the same scribbles as he reviewed the various causes of infertility, the results of the testing done two years prior, and finally, why continuing in the same treatments was unlikely to result in pregnancy. We were at the same place we'd been before.

What had changed, however, were significant advances in IVF technology. Confident because of those advances, the doctor looked me in the eye and told us he could nearly guarantee we would get pregnant with IVF. My heart skipped a beat at that new hope and I knew that we would find a way. We asked questions, took a few minutes to discuss it privately, asked more questions, and said we'd think about it. We contemplated our financial options. After all, we were still paying on our adoption loan. We prayed.

As we went through the IVF process, and in the years since, we've been selective about who we've shared details with. There's a lot of stigma surrounding IVF, especially in the Christian community. Even as we discussed this book, we debated how much to share and how much to keep to ourselves. So why am I prepared to tell you our story in depth? It is simply because of that stigma. We didn't arrive at the decision to do IVF lightly. If you're reading this while questioning the ethics of IVF, allow me to explain our reasoning. There are a couple of aspects that are common concerns.

The first concern is the ability to "play God" in various ways. Reproductive technologies have advanced significantly in the decades since IVF technology premiered in the 1970s. Pregenetic screening (PGS) was introduced in the early 1990s. As part of the initial IVF process, cells from the days old embryo are tested. Embryologists can report genetic abnormalities, sex, and assign a "grade" based on the quality and appearance of the embryo. What does this mean for couples undergoing IVF? It means that couples have choices such as what sex to choose, to use or discard an embryo with genetic abnormalities knowing it could result in miscarriage or lifelong disabilities such as Downs Syndrome, and which embryos to transfer first to have the best chance of pregnancy. Is this playing God or is this looking for the best chance at a normal pregnancy and baby? Because we decided PGS is too much like playing God and because we wanted to treat our embryonic offspring equally, we chose not to have PGS done.

The second dilemma we grappled with was what to do with any remaining embryos. The number of eggs retrieved during a given cycle can vary greatly based on the medications given, the age of the woman, and her response to the medication. Anywhere from a few to twenty or more eggs can be retrieved. Not all will be mature, meaning they are able to be fertilized. Of those mature eggs not all will fertilize and not all of those that do will continue growing. Since there's no way to determine these results beforehand, it leaves a lot of unknowns. The initial reaction is to fertilize all of the mature eggs with the hopes of having more embryos to transfer. Based on the cost and time each cycle takes, having more embryos per cycle seems reasonable.

Our problem was simple. What if it works right away? What if we walk away pregnant after a month or two but have multiple embryos frozen in the lab? The idea of future pregnancies and biological children was exciting and we would transfer those remaining embryos in the future. But we weren't prepared for five or ten more children and allowing those embryos, our tiny offspring, to remain frozen indefinitely wasn't an option we could justify. What if pregnancy proved too difficult or we had complications that would make future pregnancies impossible or ill advised? Many couples in similar situations make the decision to discard their remaining embryos, whether through simply thawing and disposing of them or donating them for medical research. Sometimes couples donate their embryos to be anonymously given to another couple. We would not do that to our embryonic children!

We found two solutions to this dilemma: to limit the number of eggs we attempted to fertilize and, if necessary, to participate in an embryo adoption program. An embryo adoption program is similar to that of a domestic newborn adoption; we would be able to pick the couple adopting our remaining embryos. When we first considered IVF prior to Derrick's adoption I struggled with the idea of placing our remaining embryos for adoption. After years of infertility how could I live knowing that another family was raising our child? It wasn't an idea I could process. Interacting with Derrick's birth mother and becoming an adoptive mother myself changed my perspective. Yes, it would be difficult knowing that another child was being raised by someone other than me, but like Derrick's birth mother, if we came to that place it would be because we couldn't adequately care for that child ourselves. Knowing that many couples in embryo adoption programs are there because of infertility reasons, perhaps having failed multiple cycles of IVF, made me tearful with joy for the gift that they would receive in a baby to call their own.

One of the things we discussed with the fertility doctor was fertilizing a limited number of eggs. That decision during our first and only cycle resulted in three embryos. Before I tell you about those three little miracles we created, let me tell you about the IVF cycle.

In Vitro Fertilization is not for the faint of heart. We had a few tests done prior to the start of the process but most of the work took place within a two week period. A few days before this started I received several large boxes. Our dining room table was soon crowded with pills, vials of medication, syringes, needles, alcohol pads, sharps containers for disposal of needles, and detailed instructions on how to inject myself with hormones designed to stimulate egg production and control my menstrual cycle. Any woman who has undergone IVF will assure you that there were no adequate warnings that these drugs introduce temporary insanity (hot flashes), mood swings, and a plethora of other unpleasant symptoms.

When the day came to start the stimulation cycle, I consulted the calendar the IVF coordinator had extensively reviewed with me: the exact dosages, times, and methods of each drug. I drew up that first medication, swabbed my abdomen with an alcohol swab and stood there frozen for a minute as I worked up the courage to plunge the needle into my flesh. Having given countless injections to patients over the years in no way prepared me for that first stick. By the time I'd stuck myself several times a day for those two weeks it was a thoughtless act. Every other day or so I'd have bloodwork and an ultrasound. Later that day the coordinator gave updated information on doses and timing for my drugs. Finally it was time for my eggs to be harvested.

We had an hour drive to the office where the egg retrieval was to be done under anesthesia. It went well and we left knowing how many eggs had been retrieved. Later that day we got the call telling us how many had been fertilized. The IVF community has many wonderful phrases that take away the formality of standard medical terminology. Embryos are called em-babies. We had three em-babies! Two days later we got the call that one had ceased growing and was no longer alive. I was devastated at the loss but forced myself to focus on the remaining embryos. Two days later we received another phone call. Our em-babies were growing slower than normal. I feared this was it and we would walk away without a chance at pregnancy. Due to the cost another cycle wasn't an option. Fortunately, by the next day they had progressed enough to be frozen.

A month later, when my hormone levels and uterine lining were optimal for implantation, it was transfer day! There is a wonderful phrase in the IVF community. After an embryo is transferred back into the uterus and in the two week wait before a blood test to confirm or deny pregnancy, one is considered "PUPO" – pregnant until proven otherwise. This was the first time that a life we had created was inside me and it felt great! Unfortunately, two weeks later we got the call that we weren't pregnant. A few days later we had what would be our final consultation with the doctor. He said that based on the few embryos that fertilized and made it to freezing and the failed transfer, we had poor egg quality. He began discussing changes in medication for another cycle. I quickly interrupted and reminded him that we had a remaining embryo. His confused expression showed that he had already moved on. He agreed to that final transfer but it was easy to see that he didn't have any confidence that it would implant.

A few weeks later I was again PUPO (pregnant until proven otherwise). 10 days later I POAS (peed on a stick) and got my BFP (big fat positive). We were pregnant! When you've waited for so long to get pregnant each day is exciting. Pregnancy is, of course, a mixture of highs and lows. Pregnancy after IVF has its unique concerns. For the first trimester I was monitored regularly and given weekly ultrasounds. We were able to watch as our little baby grew and changed in a way most couples can't. As a precaution, based on abnormalities in some of my bloodwork, I was on injectable blood thinners twice a day for the first half of my pregnancy, leaving dark painful bruises on my abdomen. A standard injection is progesterone in oil (PIO) every evening through the first trimester. I was brave and self-sufficient during the injections in the stimulation cycle. PIO is a whole new monster, though, and Jordan was forced to get involved. Every night he drew up and injected the thick medication into my hip. I blasted the song "Overcomer" by Christian contemporary artist Mandisa as I prepared myself for the painful needle stick into sore muscles. Mandisa sang me through the pain:

> *You're an overcomer*
> *Stay in the fight 'til the final round*
> *You're not going under*
> *'Cause God is holding you right now*
> *You might be down for a moment*
> *Feeling like it's hopeless*
> *That's when he reminds you*
> *That you're an overcomer*
> *You're an overcomer*[2]

Thirty-seven weeks after transfer day we welcomed Lois.[3] My pregnancy was complicated by cholestasis, a liver condition in pregnancy. Induction ended in an emergency caesarean section. She spent a few days in the NICU but quickly grew into the vibrant, lovable child she is now. The ironic thing about endometriosis, the condition that necessitated our use of IVF, is that pregnancy can clear it out. In our case that was the result! As Lois entered her eighth month of life we rejoiced in the news that we were again pregnant. I was thirty-seven weeks pregnant and had cholestasis for the second time when we had a repeat c-section and rejoiced as Jewels was born, tiny but healthy.

Jordan and I have joked for a long time that we need to write a book about our experiences, that we know all the ways to grow a family: foster care (even though it didn't end in adoption), domestic newborn adoption, fertility treatments, and creating life the old-fashioned way. Our road was incredibly painful but beautiful at the same time. I cry for the women walking the same path: the negative pregnancy tests, the complicated process, the first pink line, the first glimpse of that little person whose eyes are yours and whose nose is her daddy's, the dreams dashed, and the dreams fulfilled. God led the way, every single day, and gave us the strength to continue each long day. Would I do it all over again? In a heartbeat! Am I glad I don't have to? Absolutely! I'm glad to have this painful period behind us and parenting ahead of us.

Chapter Notes

1. Scripture quotations taken from The Holy Bible, New International Version ® NIV ® Copyright © 1973 1978 1984 2011 by Biblica, Inc. TM Used by permission. All rights reserved worldwide.

2. Copyright © 2013 Meaux Mercy (BMI) Moody Producer Music (BMI) 9t One Songs (ASCAP) Ariose Music

   (ASCAP) Universal Music - Brentwood Benson Publ. (ASCAP) D Soul Music (ASCAP) (adm. At CapitolCMGPublishing.com) All rights reserved. Used by permission.

3. Their names have been changed to protect their privacy.

DOCTOR "A"

Chapter 4

The Story of Arturo Apolinario and Kermit Gosnell

The Department of State literally licensed Gosnell's criminally dangerous behavior. DOH gave its stamp of approval to his facility. These agencies do not deserve the public's trust. The fate of Karnamaya Mongar and countless babies with severed spinal cords is proof that people at those departments were not doing their jobs. Those charged with protecting the public must do better.[1]

Kermit Gosnell Grand Jury Report

The page I looked at said "Doctor A" on top of it. It appeared to be a schedule. It listed the names of patients, how many weeks gestation they were, and the type of abortion they were getting. This scrap of paper along with a number of papers had been retrieved from a dumpster alongside the highway in front of Atlantic Women's Medical Services in Dover, Delaware. I was analyzing these papers in 2010 as part of the pro-life effort to shut down the clinic. At the time I analyzed them, I had no idea that the Gosnell Grand Jury Report would be released in January of 2011 and would ultimately result in the shuttering of this clinic.

So who was this mysterious "Doctor A?" What I discovered at the time from my research was that abortionists often disguise their identities out of fear for their safety. This is particularly true of the small number of abortionist like Kermit Gosnell that do exceptionally grisly third trimester abortions. But Doctor A didn't appear to be doing abortions past 20 weeks. So why was he so secretive? What I didn't know at the time was that Doctor A was considered to be the supervisor of Kermit Gosnell when Gosnell started his abortions in Delaware. In reality, it is unlikely that Doctor A provided any oversight of Gosnell.[2]

To understand where my story is going, you have to start by understanding the bigger story of Kermit Gosnell and his late-term abortion network. The story begins with the first paragraph of a Philadelphia Grand Jury Report resulting in the prosecution of Kermit Gosnell and his associates. It begins this way.

> *This case is about a doctor who killed babies*
> *and endangered women. What we mean is that*
> *he regularly and illegally delivered live, viable,*

*babies in the third trimester of pregnancy –
and then murdered these newborns by
severing their spinal cords with scissors. The
medical practice by which he carried out this
business was a filthy fraud in which he
overdosed his patients with dangerous drugs,
spread venereal disease among them with
infected instruments, perforated their wombs
and bowels – and, on at least two occasions,
caused their deaths. Over the years, many
people came to know that something was going
on here. But no one put a stop to it.*[3]

What we now know is that Gosnell was engaged in an interstate abortion scheme to perform illegal late-term abortions while avoiding prosecution. What made the scheme possible was the use of a drug called Digoxin and similar drugs that are used commonly by the abortion industry as a lethal injection. This drug, like so many items in the abortion industry, was not designed to kill fetuses. It was developed to treat heart conditions in adults. But the abortion industry learned that it could be used to cause a fatal heart attack in the fetus. What especially helped Gosnell and other late-term abortionists who use this scheme is that the Digoxin doesn't always kill right away. This is especially true if the abortionist misses the fetus with the injection, and injects the toxin into the amniotic fluid instead. The Digoxin can take up to twenty four hours to kill the fetus, and in some cases fails to kill the fetus. The abortion industry typically requires the patient to come back the next day to confirm that the fetus is dead before finishing the abortion. Failure to confirm death can result in viable babies born alive.

This brings us to the legal scheme. In order to charge someone with murder, you must be able to prove the jurisdiction in which the murder occurred. For example, officials in the State of Delaware can't charge you with murder if the murder occurred in Maryland. Maryland officials would have to do the prosecution. What Gosnell was doing was killing the babies after the legal limit. The legal limit in Delaware at the time was 20 weeks gestation and in Pennsylvania it was 24 weeks gestation. Gosnell was routinely killing babies capable of surviving outside the womb after 24 weeks. Here is the catch. Because Digoxin can take twenty four hours to kill the baby, prosecutors can't prove in which state the babies died!

This interstate abortion scheme isn't new. Gosnell may have learned about it from another late-term abortionist, Stephen Chase Brigham. Brigham started his illegal abortions in New Jersey, sent his patients in a procession of cars, and finish them all at a Maryland clinic. He had a secret late-term abortion facility in a store front in Elkton, Maryland. There was no signage to indicate that it was an abortion clinic. The only reason it was discovered was because another abortionist that he supervised perforated the uterus and bowel of a patient, sending her to the emergency room. The emergency room visit resulted in authorities raiding the clinic and recovering the frozen bodies of once viable fetuses. Brigham and his fellow abortionist were charged with murder in Maryland.[4]

This is where things went terribly wrong. After charging them with murder, the prosecutor later had to drop the charges. Why? Because they couldn't prove that the babies died in Maryland.[5] Brigham somehow managed to dodge the biggest legal bullet of his career. Kermit Gosnell wasn't so lucky. Unlike Brigham, Gosnell got sloppy. Gosnell wasn't confirming the death of the babies before he induced labor. He was inducing labor in women whose babies were still very much alive. This wasn't something that just happened on a rare occasion. He induced living, viable babies over and over again.

After a baby was born alive, he inserted a scissors into the back of the neck and cut the spinal cord. Because prosecutors could prove that the babies were killed in Pennsylvania after the 24 week legal limit for abortions, the murder charges stuck. If Gosnell hadn't been sloppy, he would have confirmed death before inducing labor. If the babies were still alive, he would have injected them with Digoxin again. And he would have literally gotten away with murder just as Brigham had. If you think that I am exaggerating the horror, you can read the Grand Jury Report for yourself. It is even more horrific than I describe.

It's important to understand that the killings Gosnell performed and for which he received murder charges are not substantially different than the killings performed by celebrated late-term abortionists around the country. As of the writing of this book, there are seven states that have no gestational limit on the killing of viable babies in the third trimester: Alaska, Colorado, New Hampshire, New Jersey, New Mexico, Oregon, and Vermont.[6] In these states, abortionists almost universally use Digoxin or a similar lethal injection to kill the child. After this the body is removed either by inducing labor, removing the body piece by piece, or through a partial birth abortion type procedure. We know that the lethal injection can take hours to kill the child and that if it fails, a second injection must be administered. On the other hand, Gosnell's technique of snipping the spinal cord with a scissors is quick and final. It's entirely possible that Gosnell was causing less pain and suffering to the fetus than an induced heart attack. This begs the question as to who is worse, the abortionist who is spending life in prison because he used a scissors or the few legal late-term abortionists who use Digoxin and are celebrated by the abortion industry?

Of course the abortion industry aggressively rejects any comparison between Gosnell and their late-term abortionists. Their claim is that Gosnell was entirely different because he killed them after they were born. According to the abortion industry, killing it in the uterus is entirely different than killing it outside the uterus. If you kill it after it leaves the uterus, then you are a murderer. But if you kill the exact same fetus inside the uterus, then you are a celebrated abortionist. You are a hero. The idea that this is a substantial difference is silly on its face. But I will put forward a rhetorical question for those who think that location inside or outside of the uterus determines whether you can kill it. What if one of these celebrated abortionists were to do a cesarean section, but snip the spinal cord with the scissors before lifting the baby out of the uterus? The baby is still inside the uterus. The uterus is opened up to remove the baby. But you simply snip it before lifting the baby out and cutting the umbilical cord. Is this abortionist still celebrated? Or is he now a monstrous murderer?

Take the example a little further. The doctor lifts the baby out of the uterus. Upon seeing the fetus, the mother decides that she does not want it after all. The doctor then places the fetus back inside the uterus. Once inside, he snips it's neck and removes it again, but this time dead. Is he still a celebrated abortionist? You see, there really isn't a substantial difference between the killing Gosnell did and the killing done by other late-term abortionists.

The abortion industry also claims that Gosnell was different due to the unsafe and unsanitary conditions of his clinic. It's true that the conditions were horrific. It's true that his staff weren't trained or qualified. It's true that Gosnell put women's lives at risk every day that he operated that clinic. But that's not the reason he is spending life in jail. That's not the reason he faced the death penalty. We don't charge doctors with murder because of unsanitary conditions. Even if his clinic was safe and sanitary, he still could have faced the same murder charges. The conditions don't change the fact that the manner in which Gosnell aborted these babies was not substantially different than that of other late-term abortionists.

This brings us back to "Doctor A." The abortion clinic staff were very careful to not write down his real name. But someone made a mistake. Someone in that clinic threw part of a handwritten prescription into the trash. As I sorted through these scraps of papers, this

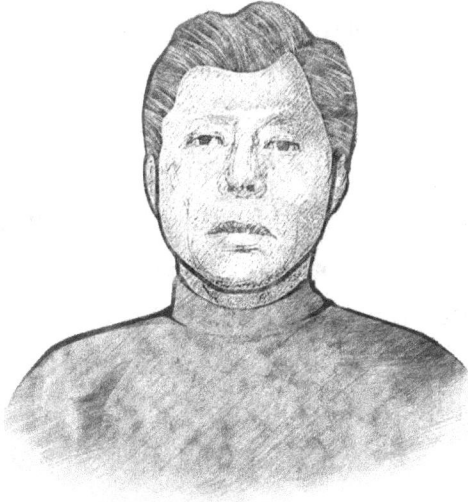

Arturo Apolinario, MD

little slip of paper went through my hands and I got our big break. There on the cancelled prescription was the name Dr. Arturo Apolinario, MD.

Once we knew the name, I was able to go to work researching him. Apolinario was born in 1936 in the Philippines, a beautiful Asian country comprised of over 8,000 islands. He immigrated to the United States where he died in 2016 at the age of 80.[7] He only retired in 2011 at the age of 75 after the State of Delaware suspended his medical license in the wake of the Gosnell Grand Jury Report. For decades he lived and practiced medicine at hospitals in Philadelphia. That all changed in 1993 when he applied to practice medicine in Delaware and went to work as an abortionist at the age of 60.[8] He performed abortions at Atlantic Women's Medical Services in Dover, Delaware. They were only open on Fridays. And so he commuted from Philadelphia for a full day of on average thirty abortions. The clinic was generally open from 9:00am till 3:00pm. A typical six hour day with 30 patients meant that there was only 12 minutes per patient to do abortions. Unfortunately, these packed schedules are not unusual in abortion clinics. When you are doing rushed back to back surgeries, it is reasonable to expect a higher rate of complications.

His packed schedule resulted in a lot of income. My conservative estimates had him generating at least $12,000 in gross income for the clinic each day that they were open. This totals over $600,000 annually for this clinic which was only open one day per week. This does not include income from the sale or donation of fetal body parts. I have no way to estimate that.

At the time, the shocking undercover Planned Parenthood body parts videos had not yet been released. But it was well known in the pro-life community that selling body parts was something that some clinics did. When I started researching Atlantic Women's Medical Services, the possibility that they were selling body parts didn't cross my mind. That all changed when an employee of a local cable company approached a well-known local pro-life leader in our community. The cable guy was distraught. He was working in the clinic and happened to come across a freezer in the basement. Out of curiosity he opened the freezer and saw all the bloody parts of babies. A clinic employee caught him looking in the freezer and yelled at him. He freaked out. Of course I would be very skeptical if someone told me this. Did he really understand what he was looking at? Did he make it up? But the details of his story made it credible. The basement had windows that made it very easy to see inside. I had already known that there was a freezer in the basement. And I had already known that there was a lady that would occasionally come on abortion days with a cooler that she would struggle to carry from the clinic to the trunk of her car. So I found the cable guy's story to be very believable.

The cable guy had already contacted authorities, but they weren't interested in doing anything about it. It is legal in the United States to possess the bodies, body parts, and organs of unborn children and to give those body parts to another person for free. It is not legal, however, to buy and sell the body parts. The problem with making it illegal to buy and sell body parts is that it is very difficult to enforce. We didn't have any idea what financial arrangement Atlantic Women's Medical Services had with the woman who took the body parts. The problem is that cash is fungible. The word "fungible" simply means that cash isn't attached to any specific good or service. This is why it is so easy to hide financial arrangements with cash. For example, you could hide your income from the IRS. Or you could buy illegal drugs with cash so that no one can prove that you bought the drugs. The same is true of buying fetal body parts. You could claim on paper that you are gifting me body parts. But at the same time you could rent me an overpriced room in the clinic. And no one would know that you are selling me body parts because the arrangement is done with a hand shake, not on paper. The only way to stop this sort of illegal body parts trafficking by abortion clinics would be to entirely ban or highly regulate the donation of body parts as well as the sale of them.

Trafficking fetal body parts wasn't the only thing sketchy about this clinic. There were also the numerous county safety code violations. While researching the clinic, I discovered insufficient parking spots, a major safety violation. The paved parking lot only had room for about five or six cars. Three of those spots were typically used by abortion clinic staff. The staff took the spots closest to the doors for safety reasons. This left two or three parking spots for approximately thirty patients. It was also a few spots less than the county required. Most of the patients were parking at nearby businesses. What really shocked me and caused me to file a complaint with the county was the phone call I made to the clinic asking them where I should park. The clinic is along a four lane major highway. The lady who answered the phone told me to park on the shoulder of the highway. I was shocked that the clinic would be so careless as to tell patients to park along the shoulder of a busy highway. The highway department had even put up a sign on the shoulder saying "NO STOPPING STANDING OR PARKING."

The county conducted their investigation and found a number of violations. The inspector even called me to confirm that they did not have enough parking spots. He also told me that they had numerous violations. Unfortunately, I was not able to get my hands on the inspection report. But the upgrades were apparent to the public. They were required to add parking spots to bring it up to code. They dumped some stones to create two more parking spots resulting in at most five parking spots for thirty patients. The second change was that they replaced the old wood burning furnace with a new electric heat pump. Wood burning fireplaces are a fire hazard which caused the county to require it to be replaced. The condition of the building was enough to concern anyone. It did not look like a legitimate clinic. In fact it reminded me of what I'd assumed a so-called "back alley clinic" would look like. Only it wasn't illegal and it wasn't in a back alley. The building was obviously built to be a residential home. It was a basic ranch style house on an unfinished basement. It was on a major highway but it butted up against a residential neighborhood. The building had not been well maintained. No one would want to actually live in this house. All the first floor windows had either been filled in and covered with siding or had been replaced with small high glass blocks to let in light. The only normal windows were the upstairs windows and the windows that allowed a clear view into the basement. The front doors featured rust. One door was for patients and the other was for staff. The mailbox was at an angle and the spots on the house with original wood siding had badly peeling paint. During the winter you could see the smoke from the wood fireplace billowing out the chimney. Clinic staff had to start and monitor a fire in this clinic while doing abortions.

Most people don't realize that these dilapidated abortion clinics that remind us of so-called "back alley abortions" still exist in the United States. Planned Parenthood does a very good job of using relatively nice-looking buildings to put a good face on the abortion industry. When most people think of abortion, they think of relatively attractive Planned Parenthood clinics, not the old dilapidated buildings used by independent abortionists. It's important to understand that there was a vast underground abortion industry before it was legalized. When abortion was legalized, the underground industry didn't disappear. It simply became legal. The same bad actors and the same bad practices from before legalization continued after legalization. This makes abortion unique from the broader health care industry. I'm not aware of any other health care industry in the United States that began as a vast illegal industry and then became legalized overnight. The uniqueness of the abortion industry is one of the reasons why abortions should be regulated differently than surgery in general. Abortion has a unique history of illegality. The only way to rid the abortion industry of the unsafe and unsanitary practices that have stuck around for decades is to force it to reform through stricter regulations and inspections. Unfortunately, the abortion industry has become a powerful lobby. They spend tens of millions of dollars each election cycle to ensure that they are not regulated and reformed. It is this climate of non-regulation and willful ignorance on the part of authorities that has allowed the bad actors like Gosnell, Apolinario, and numerous others to thrive.

In March of 2011, all my hard work paid off as the State of Delaware suspended Apolinario's license, forcing him into retirement. In order to get the State to investigate Apolinario, I, along with Nicole Theis, president of Delaware Family Policy Council, had to execute a well planned press conference. The press conference was held on the sidewalk outside of Atlantic Women's Medical Services with two clear goals in mind. We wanted the State of Delaware to investigate Dr. Apolinario and we wanted the world to see this dilapidated building. At this point, Apolinario wasn't anywhere on the media's radar. So I teased the media with the promise of revealing his name. It worked. The media showed up, including regional TV stations.

Nicole delivered her pre-written speech with the clinic in the background. Our big reveal was that Apolinario's controlled substance license had lapsed. He was illegally using drugs for the abortions he performed without the proper controlled substance license. We pointed out that the investigation into Gosnell also started as a controlled substance investigation. We connected Apolinario to Gosnell and demanded an investigation into Apolinario. Normally a lapsed controlled substance license wouldn't warrant much attention. But due to Gosnell, it got a lot of attention. One local TV station in particular even took close up shots of the various dilapidated features of the building. I couldn't imagine that we would be so successful. A few days later the Attorney General's office confirmed that they were investigating Apolinario. Not long after the press conference, the State announced that they were suspending his license. Apolinario would never practice medicine again.

Chapter Notes

1. Williams, R. Seth. *Report of the Grand Jury*. Court of Common Pleas for the 1st Judicial District of Pennsylvania (2011) https://cdn.cnsnews.com/documents/Gosnell,%20Grand%20Jury%20Report.pdf

2. Delaware Secretary of State "State Suspends Second Physician, Dr. Arturo Apolinario, with Ties to Gosnell" *Press Release* (2011). https://sos.delaware.gov/newsroom/state-suspends-dr-arturo-apolinario/

3. Williams

4. Press, Eyal. "A Botched Operation" *The New Yorker* (2014) https://www.newyorker.com/magazine/2014/02/03/a-botched- operation

5. Kilar, Steve. "Unclear if fetal deaths occurred in Md., Cecil prosecutor says" *The Baltimore Sun* (2012) http://www.baltimoresun.com/news/breaking/bs-md-fetal-case-dropped-follow-20120307-story.html

6. "States with Gestational Limits for Abortion" *Kaiser Family Foundation*

   https://www.kff.org/womens-health-policy/state-indicator/later-term-abortions/

7. "A Wetzel and Son Obituary for: Arturo Nelson Alcantara Apolinaro, MD" (2016) http://www.wetzelandson.com/browse-record php? recid=26585

8. State of Delaware "Application for License to Practice Medicine and Surgery"(1993)

   Appolinario's application to practice medicine was obtained via a FOIA request. The illustration of Appolinario is taken from his picture in the application.

Part B

EMBRYO & FETAL DEVELOPMENT

Fetal Beauty

THE BEGINNING OF YOU

Chapter 5

Our Biological Origins
Through Conception

Where do you come from? You could answer that question a lot of ways. Spiritually, I come from a Christian home. More specifically, I come from an ethnically and religiously Mennonite home. Geographically, I come from Sussex County, Delaware. I was born and raised between Milford and Greenwood. And now I am raising my family here. Familially, I am a Warfel on my father's side and a Mast on my mother's side. But in this chapter we are going to answer that question biologically. What are your actual physical origins? Where did you materially come from? The answer may amaze you. It's almost too fantastic to be true.

If you are like most people, the furthest you have answered this question is to identify your biological parents. If you are very fortunate, you have a relationship with your biological parents. You share their DNA. It is the biological tie that binds. Of course family is much more than biology. I should know since I adopted a son who is every bit a Warfel as my biological girls. Maybe you aren't as fortunate as me. Maybe you don't have a relationship with one or both of your biological parents. Maybe you don't even know who they are. Maybe you have nagging questions. Biology plays a big role in our identities.

What we are going to look at in this chapter is the very beginnings of you. Sure, there was a sperm and an egg and the rest is history. But where did those cells come from? How far back can we trace them? Luckily for us, modern biology has a lot of answers! Biology also has an unsolved mystery, the mystery of you and from where you actually came.

The descriptions I give in this book are a vast simplification of the biology in order to give a basic understanding of where you came from. The actual biological processes are far more intricate and awe inspiring than I am capable of describing. You can learn more about the biology by picking up an embryology book such as *Larsen's Human Embryology*.

According to embryologists (embryology being a sub-discipline of biology), the very beginning of you can be traced back to your parents when they were embryos. The very first two cells, one from your mother and one from your father, that eventually become you are called primordial germ cells. For this book I will simply call them germ cells. These cells first appear about four to six weeks gestation in your embryonic parents.[1] They first appear in the yolk sack of the embryo, not the embryo proper. The embryo proper is the part of the embryo that actually becomes the body of the fetus. A few weeks later these germ cells actually migrate or move from the yolk sack to the genital region of your parents' tiny embryonic bodies. The same was true of you when you were an embryo. The germ cells that will or have become your children also formed four to six weeks into your embryonic life. Many women haven't even taken a pregnancy test at the point when these germ cells first form in the embryo. But their embryonic offspring already have the cells to be their grandchildren at this very early stage. While we don't exactly know from where these germ cells come, it is believed that they may form in the first two to three layers of cells of the embryo proper.[2] Where these germ cells first come from is one of the mysteries of life. Science may someday give us a better explanation, but for now we are left only to reflect and marvel at this mystery and the intricate series of events that lead to our existence.

These powerful little germ cells have the potential to become many different kinds of tissue. While their purpose is to become the sperm and eggs for the next generation, sometimes a stray germ cell may end up migrating to where it doesn't belong. When this happens, it could become a bizarre tumor called a teratoma that later has to be removed. Because these germs cells have the ability to become many different kinds of tissue, the results are some mind blowing tumors. These tumors may include teeth, hair, and even the odd eyeball.[3]

Once these germ cells reach the genital region of the embryo, they divide a few times, resulting in many more germ cells. As they divide, they each have a complete set of 46 identical chromosomes. The chromosomes are organized into 23 pairs. The process of cells dividing into two new cells is called mitosis. This is how most living things grow. After the germ cells in the embryo divide a few times, the germ cells arrest. This means that they cease to divide. The future of these germ cells is different in boys and girls. The germ cells arrest in both the boy embryo and the girl embryo. But the process of turning these cells into eggs in girls and sperm in boys is very different.[4]

In order to become sperm and eggs, the germs cells in both boys and girls divide. But they go through a very different kind of cellular division called meiosis. This is a different kind of cellular division than the mitosis we discussed earlier. Instead of the germ cells dividing into two new identical cells as is done in mitosis, the germ cells divide in such a way as to result in four new cells. Each of the four new cells only have half of the 46 chromosomes that were in the original germ cell. Chromosomes in humans are organized into 23 pairs. But the new sperm and egg cells resulting from meiosis have only half of the pairs.

The difference between boys and girls is that meiosis is done differently. In boys, the germ cells are arrested in the embryo before they even start the process of meiosis. They remain arrested until puberty. At puberty some of the germ cells continue to divide resulting in a lifelong supply of germ cells. But some of the germ cells go through meiosis. Each germ cell going through meiosis results in four new sperm cells with the chromosomes halved. Each of these sperm cells only have 23 of the 46 chromosomes. The male will be able to continue producing sperm cells through the rest of his life, barring a medical condition to cause him to become infertile.

In girls, the germ cells begin meiosis while they are still embryos in the womb. The germ cells begin dividing to produce eggs but go dormant before finishing. They remain dormant in the ovaries until puberty. When the girl reaches puberty, ovulation begins. Each month, meiosis resumes for one of the germ cells resulting in an unfertilized egg. While the germ cell results in four new cells, only one of the cells develops into an egg that is ready to be fertilized. This egg has only half the chromosomes as well. At this point meiosis freezes yet again. The process of meiosis won't actually be completed unless the egg is fertilized. Mitosis, the process by which most cells divide and grow, no longer occurs in a female's germ cells after puberty. This means that she has a limited number of germ cells. And each germ cell can only result in one unfertilized egg. That is why girls have a limited number of eggs while boys continue to have sperm as long as they are alive.[5]

As you will see in the next chapter, you do not yet exist as a new living human in this order of events. The cells that will become you, the egg from your mother and the sperm from your father, have not yet formed the new living organism that is you. That won't occur until the egg and the sperm fuse together in the process that most people know as conception. Chapter 6 will explain what a living thing is, when a new life begins, and why biology has definitively answered the question of when life begins.

Now that we know the origin of the sperm and the egg, the next stop is conception!

Normally each month, a woman of child bearing age will ovulate or produce a mature egg from one side of her ovaries. The egg is ovulated out of a follicle in the

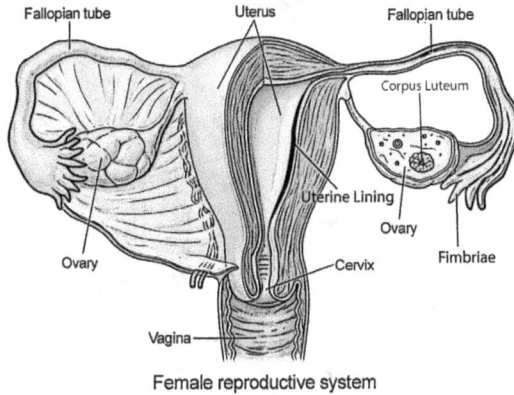

Female reproductive system

Female Reproductive System

ovary. The unfertilized egg is then stuck to the side of the ovary. The follicle the egg was ovulated from does something extraordinary. It turns into a hormone producing gland called the corpus luteum. The corpus luteum secretes the hormones, especially progesterone, which are necessary to support the lining of the uterus. The lining is kept nice and supportive for the embryo that could result if the woman is sexually active.[6]

At this point, finger like structures on the end of the fallopian tube take action. These fingers are called fimbriae. The fimbriae scrape the egg off of the ovary and guide it into the fallopian tube. If the woman is sexually active, this is where the egg meets the sperm to be fertilized in a highly intricate process that I can only describe as a sophisticated chromosomal dance.

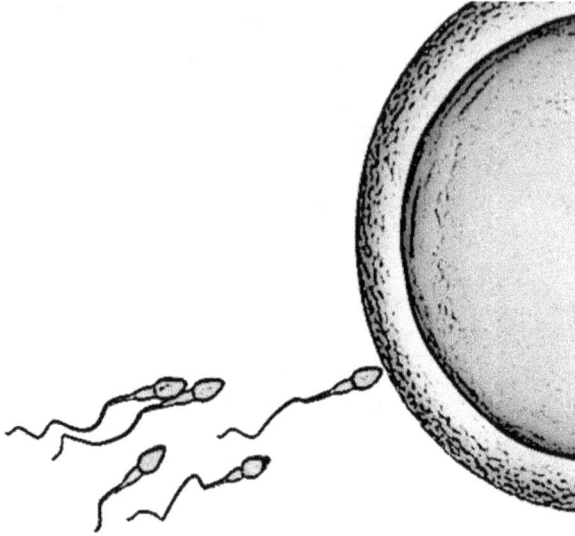

Sperm Attracted to Egg

Once the sperm cells and the egg are in the fallopian tube, they are ready for fertilization. Typically there might be hundreds of sperm waiting for an egg to fertilize. When the egg enters the fallopian tube, for some unknown reason, mature sperm cells are attracted to the egg. They swim to the egg and try to burrow their way in.[7] Interestingly, research now shows that the egg may be somewhat picky about the sperm that it attracts based on the chromosomes that are desired. It is almost as if the egg gives a "come hither" look to the sperm cells that she finds attractive. We don't actually know why the egg appears to attract certain sperm and the research was done on mice not human cells. But what we do know is that the more we research, the more we find that the egg is playing a larger role in the process than previously thought.[8]

Now that there are numerous sperm attempting to burrow into the egg, the challenge is for the egg to only be fertilized by one sperm. This challenge is accomplished through the use of a hard shell and chemical switches. The hard shell of the egg keeps the sperm out. Around the shell are chemical switches waiting to be activated by a sperm cell. When a lucky sperm cell hits one of those switches, for a brief moment the shell softens, allowing the sperm to enter the egg. As the sperm enters the egg, another switch is activated that causes the shell to harden again so that it is no longer possible for a sperm to enter.

In 2016, researchers at Northwestern University were able to capture images for the first time of a biological occurrence during conception called the zinc spark. These researchers called this exciting new discovery "radiant zinc fireworks." While the existence of the zinc spark was already known, these scientists were able to capture images of the zinc emanating from human eggs for the first time. The zinc spark is part of the biological process of the sperm entering the egg to fertilize it. As the sperm enters the egg, chemical changes in the egg result in excess zinc that is no longer needed. The excess zinc is emitted from the egg in a flash. What these researchers were able to do with the help of technology was to capture the zinc in the form of visible light in what you might call a fireworks show. The scientists used sperm enzymes to simulate fertilization and cause the zinc flash without actually creating a new living human. You can view a video of the zinc flash published by Northwestern University by following the link in the chapter notes.[9]

Now that the sperm cell has entered the egg, the next step is for the chromosomes of the two cells to fuse together, creating a new and complete set of chromosomes. Each cell brings half of the 23 pairs of chromosomes. The chromosomes from the egg and the sperm line up opposite of each other in something like a sophisticated dance. The halved chromosomes from each parent duplicate themselves and fuse together into a completely unique, new set of chromosomes. The egg divides into two new cells with 23 complete pairs of chromosomes in each of the two new cells of the new embryo.[10]

The fusing together of the chromosomes from the two parents is something biologists call reproduction with heredity. The DNA in your chromosomes contains all of your characteristics. It is the blueprint for your body. One of the key defining characteristics of a living thing is that it passes on its characteristics through DNA to the next generation. Living things pass on their

DNA in a way that allows them to change, adapt, and evolve over time. In humans, conception and the fusing together of the chromosomes from both parents is what makes this possible. In passing on your DNA to your children, they will have characteristics from you and the other parent, but they will not be entirely like either of you. Each generation changes and is different from its ancestors. This is reproduction with heredity. You will learn more about this in the next chapter.

One of the characteristics passed down to the child through chromosomes is the chromosomal sex of the child. In today's ideological controversies, gender has taken on a whole new meaning. This is not a book for or against any gender ideology. I'm simply referring to the biological and genetic sex of the child. This characteristic is determined by the 23rd pair of chromosomes called the sex chromosomes. This pair can be made up of either two 'X' chromosomes or an 'X' and a 'Y' chromosome. A female set of chromosomes will have two 'X' chromosomes and a male will have one of each. During conception, the egg provides one 'X' from the mother. The sperm could provide either an 'X' or a 'Y'. Whichever is provided by the sperm will determine the gender of the new child.

Sometimes during the process of the creation of egg and sperm cells and the process of conception, something will go wrong with the chromosomes. The new embryo may not have a correct set of 23 pairs. The abortion industry and many people in health care refer to this as "chromosomal abnormalities." I don't typically use the word abnormal because it is dehumanizing and stigmatizing. We don't talk about adults with Down syndrome as abnormal humans. But the only difference between a fetus with Down syndrome and an adult with Down syndrome is time and development. I'm especially encouraged by the recent trend to refer to people with disabilities as "differently abled." People with disabilities may be different, but they have skills, abilities, and passions like the rest of us. I don't believe it's moral to stigmatize adults with disabilities. And likewise, I don't believe it's moral to stigmatize unborn children with disabilities by referring to them as abnormalities.

The genetic disability with which most people are familiar is Down syndrome. This is a condition where a third #21 chromosome is attached somewhere that it does not belong. Human chromosomes are supposed to come in pairs. But when a third chromosome is tagged onto the end of a pair, we call that a trisomy. And so Down syndrome is trisomy 21. There are many other trisomies. Another common trisomy is #18. In chapter 1 you read about my experience parenting twin foster girls with trisomy 18. For many embryos with chromosomal disabilities, the condition is terminal.

Many fail to implant, miscarry, or are still born. But then there are those that beat the odds.

One such person who beat the odds is John Stoklosa from Newark, Delaware. Mr. Stoklosa is a competitive powerlifter and is among the elite. Down syndrome hasn't keep him from following his passion. He doesn't just compete in the Special Olympics where he won gold. Mr. Stoklosa competes in regular events against other elite athletes who do not have disabilities. He has benched over 400 pounds and earned the respect of his peers.[11] Unfortunately, John Stoklosa isn't the image most people have in their minds when their unborn child receives a diagnosis of trisomy. These unborn children are more likely to be aborted than carried to birth.

After conception, other things can go wrong as well. Normally the newly fertilized egg will make its way down the fallopian tube to implant in the uterus. But sometimes things don't go the way they are supposed to and the embryo implants inside the fallopian tube. In rare cases, the egg may not even make it into the fallopian tube and implant in the abdomen. These pregnancies are called ectopic pregnancies and they are always life threatening for the mother. The most serious concern is that the embryo will cause blood vessels to burst, causing massive and even deadly internal bleeding. The two treatments typically used to save the mother's life are both abortions in the broad sense of the word abortion. One is the use of an abortion causing drug very early in the pregnancy to kill the embryo. The other is a surgery to remove the embryo intact.

In terms of abortion, it is important to understand that treatments for ectopic pregnancy are not the kinds of abortions done in abortion clinics. If you go to your local abortion clinic for an ectopic pregnancy, they will not treat you. They will send you to the emergency room. The abortion procedures done by abortion clinics are not treatments for ectopic pregnancy. It's also important to understand that not all abortion clinics take the necessary care to confirm that you do not have an ectopic pregnancy before doing an abortion. Any abortionist who cares about his patients should be doing ultrasounds before the abortion to check for complications. If your embryo is in the uterus, an ultrasound should confirm it. But if no ultrasound is done and the abortionist proceeds with an abortion, the abortionist could accidently cause the blood vessels around the embryo in the fallopian tube to burst, putting the woman's life in danger.

For some pro-life people, treating an ectopic pregnancy with a surgery is considered more ethical than an abortion causing drug. The reason is that surgery causes the embryo to die indirectly as opposed to a medication that kills the embryo directly. In the surgery, the embryo is removed intact. Assuming the embryo is alive, it will remain alive for a period of time after being removed. These surgeries give us a unique glimpse into the humanity of these tiny human children after they are removed and observed moving or swimming in the final moments of their lives. With the surgery, the embryo dies as a result of being deprived of the mother's oxygen and nutrients that he is no longer receiving through the umbilical cord and placenta. Regardless of which treatment you choose, the pro-life community is in universal agreement that these incredibly tragic ectopic pregnancies are sufficient justification for intentionally ending the life of the child. When pro-life people support an abortion exception for the life of the mother, these are the kinds of tragic situations that we have in mind.

I hope that you now have a greater appreciation for the complex sequence of events that led to your life and a better understanding of your biological origins. There is nothing simple about the beginning of human life. As you will read in the next chapter, there are many mysteries about the beginning of life. But there is no mystery about when life begins. Biology has answered that question definitively.

Chapter Notes

1. Schoenwolf, Gary C. , et al. *Larsen's Human Embryology*, 4th *edition*. Philadelphia: Churchill Livingston, 2009, pp. 15

2. Ibid, pp. 19

3. Ibid, pp. 18

4. Ibid, pp. 20

5. Ibid, pp. 24

6. Ibid, pp. 37

7. Ibid, pp. 39

8. Weston, Phoebe. "Fussy eggs actively choose sperm with the best eggs suggesting that fertilization is NOT random" *Daily Mail* (2017)

    https://www.dailymail.co.uk/sciencetech/article-5092821/Fussy-eggs-choose-sperms-best-genes.html

9. Paul, Marla "Radiant Zinc Fireworks Reveal Quality of Human Egg" *Northwestern University* (2016)

    https://news.northwestern.edu/stories/2016/04/radiant-zinc-fireworks-reveal-quality-of-human-egg

10. Schoenwolf, pp. 40

11. "Meet the Down's Syndrome man that's become an elite athlete – regularly out lifting his unimpaired competitor" *Daily Mail* (2013)

    https://www.dailymail.co.uk/news/article-2407982/Meet-Downs-Syndrome-man-thats-elite-athlete--regularly-lifting-non-impaired-competitors.html

NEEDLESSLY CONFUSED

Chapter 6

How We Know When Life Begins

That's a question medical folks have dealt with, and I'm not a doctor. I've spent a lot of time with ob-gyns, and they will tell you there is no specific moment when life begins.

Cecile Richards[1]

One of the nagging problems of the endless abortion debate is the needless confusion over when life begins. Americans are deeply conflicted and have been so for decades in part because of this confusion. The assertion of the mainline pro-choice movement is that you can't know when life begins. The reason I call this confusion needless is because we certainly can and do know when life begins. There is no need to speculate, ponder, or agonize about this question. We simply need to ask the experts. So who are the experts on when life begins? This is where much of the confusion lies.

The world is full of experts on just about everything imaginable. But not any expert will do. Let's say that I want to know tomorrow's weather forecast. Would I ask my medical doctor? Or should I ask my pastor? What about my attorney? These people are all experts in their fields. They are all smart people. But none of them are experts on the weather. No, I'd have to ask my local meteorologist. If you want to know when life begins, you have to ask the experts in the field of the study of living things. We call this field biology. Unfortunately, many people are needlessly confused because they look to the wrong experts. Philosophers and religious leaders are not experts in the study of living things. They are experts in philosophy and theology respectively. But they can't answer the question of when life begins any more authoritatively than your local meteorologist or attorney. You have to turn to biology if you want to understand living things.

Biology is the branch of science that studies living things or living organisms. For biologists to know the scope of their field of study, they must first define what living things are. After defining what are and are not living things, biologists are able to know what they are and are not studying.

When studying biology in high school or college, one of the very first things covered in the textbook is the definition of living things. Biologists all agree that the way we know what are living things and what are not living things is by the way they behave. We can identify living things because they behave like living things. So how do living things behave? Biologists have studied and categorized behaviors or characteristics that describe living things. These lists of characteristics are shared by living things and are the definition of what is or is not a living thing. For the purposes of this book, I'll simply refer to these as the characteristics of life. Different biologists list, describe, and organize these characteristics in different ways. Some list only four or five groups of characteristics while others break them down into seven or more groups. But regardless of whose list you use, they are all describing the same basic characteristics that are shared by living things. You will come to the same conclusions regardless of which list you use. For the purposes of this book, I will be using *Biology* 4th *edition* by Raven and Johnson, a commonly used textbook for AP classes in public high schools. You can use your own biology textbook and see the same basic unifying characteristics.

Raven and Johnson define living things with the following two statements:

Definition of life *Living things share the following basic characteristics: a degree of orderliness; the ability to respond to stimuli; the capacity to grow, develop, and reproduce using hereditary molecules; and the possession* *of regulatory processes that control and coordinate life functions.*[2]

All living things on earth are characterized by cellular organization, growth, reproduction, and heredity. These characteristics serve to define the term "life." Other properties that are commonly exhibited by living organisms include movement and sensitivity to stimuli.[3]

And so the four groups of characteristics that they describe and which we will explore are as follows:

1. The capacity to grow, develop, and reproduce with heredity
2. A degree of orderliness which includes cells
3. Ability to respond to stimuli and move
4. Regulatory processes

It is important to understand that none of these characteristics by themselves define life. It's only when you look at the big picture that the difference between living things and non-living things becomes clear. For example, living things generally move. Cars move. Is a car a living thing? Obviously it is not. That one characteristic of moving by itself does not define a living thing. In another example, my finger is composed of cells. Living things are composed of cells. If I cut off my finger and put it on ice to keep it viable, is it a living thing in and of itself? Obviously it is not. It is composed of cells and has some of the properties of a living thing. But it lacks many others such as the ability to reproduce other baby fingers or the ability to regulate itself. If you were to attach the finger back onto my hand and if it were still viable and able to continue being my finger, then you could say that the finger is part of a living thing which is me. But it is not a living thing in and of itself. As we look at these characteristics, keep in mind that it is the big picture of all the characteristics that tells us what are and are not living things.

Growth, development, and reproduction with heredity are the first set or characteristics we will explore since these are the clearest signs of life. While we might debate which living organisms have which sets of characteristics of life, this group of characteristics is common to all living things and relatively clear to understand. Growth and development in particular is the characteristic that has brought so many pro-choice people to understand that life begins at conception. It is important to understand that development doesn't finish or become complete at some point during pregnancy as many pro-choice people seem to believe. Development is a lifelong process that only ends when the organism dies. *Larson's Human Embryology* puts it this way:

At birth, the baby or neonate breathes on its own, but development does not cease simply because birth has occurred. Although this textbook discusses only prenatal development, it is important to remember that development is not just a prenatal experience; rather, development is a lifelong process, with aging and senescence involving further developmental events.[4]

If you were to enter any IVF clinic, you would find that they all use growth to determine whether or not an embryo continues to be alive or has died. Before transferring the embryo into the woman's uterus, they look to see if the embryo's cells are still dividing. Cellular division is growth in the embryo. If the embryo is growing, it is alive and transferred. If the embryo is not growing, it has died and is discarded. Unfertilized eggs do not grow. Sperm cells do not grow. You will never see an unfertilized egg or sperm cell grow and develop into anything. If the sperm does not fertilize the egg, those cells will simply disintegrate and disappear. But at the moment of conception, specifically when the DNA from the sperm and the egg finish fusing together, growth and development immediately begin. As soon as the DNA fuses together it immediately divides into two new cells of the new living thing, the human embryo. If nothing interferes with that fertilized egg, it will continue to grow into a fetus/baby, then a toddler, then an adolescent, and so on until it dies of old age. That is the magic of the beginning of life. The sperm and the unfertilized egg will never grow or develop into anything because they are not living things. This is the clearest way to understand that life begins at conception.

Reproduction with heredity is another characteristic of life that makes crystal clear when life begins. It is the characteristic most alluded to by the pro-life movement. For example, the March for Life's 2019 motto was "Unique from Day One." Heredity simply means that living things have characteristics encoded in DNA that are passed on to each generation, allowing them to change, evolve, and adapt over time.[5] This is true of all living things. Conception is what makes reproduction with heredity possible in the human species. As half the chromosomes from each parent fuse together, a new living human with a unique set of DNA is formed. The new human life has the characteristics of its parents, but he or she has also changed through the process of conception. This is heredity. Random tissue cannot do this. Formless blobs of cells do not do this. A sperm cell by itself cannot do this. An unfertilized egg by itself cannot do this. But a fertilized egg is different because it is a new living human. A fertilized egg is the result of reproduction with heredity, and if left to go its natural course will likely grow up to also produce children.

While growth, development, and reproduction with heredity are the clearest indications of when life begins, there are other characteristics of life as well. Another characteristic of life is metabolism. Living things take in energy and use that energy for growth and development. This is called metabolism.[6] The embryo has to take in energy from its surroundings to continue growing. We know that the embryo takes in surrounding fluid before it implants[7], is supported by the lining of the uterus when it implants[8], and then receives energy from the placenta and umbilical cord throughout the rest of pregnancy. Metabolism is another characteristic of the living human embryo.

The list of characteristics of life goes on. The fertilized egg is made up of cells which are complex and orderly. It responds to its surroundings as it moves down the fallopian tube taking in fluid and then it implants into the uterine lining. And it regulates itself through cellular division as the cells differentiate themselves. Biology has certainly answered this question of when life begins and answered it definitively.

So what do individual biologists say about abortion? It shouldn't surprise us that scientists and people in general tend to avoid letting their views be known about such a highly controversial topic.

Expressing your opinion can come with a cost. And when your livelihood depends on grants and donors, you will certainly avoid becoming overtly political. Biology textbooks simply teach the science. And that's exactly how we should want it. I don't want scientists to tell me what to believe about politics. That is partly because it's not their job and partly because I don't want science to become politicized and lose its credibility. I just want them to teach me the science. I don't want biologists to jump into politics and declare pro-lifers the winners. The truth is enough. That is precisely what biology tells us. And that is why we know with certainty when life begins.

There is a second reason that biologists are reluctant to wade into abortion. That is because the overwhelming majority are pro-choice and they know that their field doesn't support their ideological position. A doctorate dissertation published in June of 2019 surveyed over 5,000 biologists around the world about this question of when life begins. A whopping 85% of biologists identified themselves as pro-choice. And so biologists are far more likely to be pro-choice than the average population. But interestingly, they are pro-choice despite the biology, not because of it. The survey included a list of five different statements saying that life begins at conception. Each statement was worded differently from very innocuous to very pointed. A decisive majority of biologist agreed with all five statements. Even a strong majority of biologists that describe themselves as "very pro-choice" agreed with the statements. 64%-90% of very pro-choice biologists agreed that life begins at conception, depending on how the statement was worded.[9] And so it shouldn't surprise us that biologists want to avoid the abortion debate. Once they concede that life begins at conception, it is much harder to justify legalized abortion.

On occasion you will find biologists who wade into the abortion issue, such as the outspoken pro-choice biologist Ricki Lewis who holds a PhD in genetics from Indiana University. Lewis is a good example of how biologists justify being pro-choice despite the biology, not because of it. She wrote a blog entry in which she takes issue with pro-lifers like myself who cite the science.[10] How does she refute our arguments? Well, she doesn't. Instead she claims that we wrongly take "a leap off the page." What is this leap off the page? We simply take the clear and plain meaning of the science and apply it directly and objectively to the question of when life begins. Apparently this is something she thinks we ought not do. Instead she would have us dismiss her profession and focus instead on "philosophy, politics, psychology, religion, technology, and emotions." Yes, you read that right.

Lewis thinks that the question of when life begins is a question for your emotions, not biology. Lewis twice lets slip the problem with her argument. She claims to personally believe that life begins slightly before viability when the fetus has a slim chance of survival after birth. But she also refers to the embryo as a "prenatal human." And she refers to an early fetus as an "organism." Organism is, of course, synonymous with "living thing." So in other words, she believes that before life begins, it is a human and it is a living organism. How can it be alive before life begins? How can it be a human before life begins? I suppose only Ricki Lewis can answer that question.

But if the science is so clear, why is there so much confusion? And how come so many people don't believe it? The answer lies in the brilliant propaganda of the pro-choice movement. The propaganda is so powerful and so devastating that it convinced half the county and even the Supreme Court to buy into it. The pro-choice movement knows they don't need to convince us to adopt their position on when life begins. They only need to sow enough confusion to convince people to believe it is an open question. The way they did that is by appealing to every unrelated expert, but not to science. Primarily they pointed to religion and philosophy broadly as an alternative to science. The difference is that biology tells us when life begins while philosophy and religion answer questions that science can't answer about meaning and purpose in life. These aren't the same questions but they are similar enough that the pro-choice movement was successful in convincing people to look to philosophy and religion instead of to science. Of course philosophy and religion can't tell us when life begins. Only biology can do that. And so the widely held position of the pro-choice movement is that you can't know when life begins.

Unfortunately, this propaganda was so successful so as to even dupe the Supreme Court. In Roe v Wade, which legalized elective abortions in the entire country, Justice Blackmun for the majority wrote:

> We need not resolve the difficult question of when life begins. When those trained in the respective disciplines of medicine, philosophy, and theology are unable to arrive at any consensus, the judiciary, at this point in the development of man's knowledge, is not in a position to speculate as to the answer.[11]

Notice how Justice Blackman conveniently fails to mention biology or even science broadly in this opinion. Instead he mentions medicine, philosophy, and religion. (The practice of medicine is informed by science but a medical doctor is not typically a scientist.) Here we see that the entirety of Roe v Wade and abortion on demand in America hinges on a failure to ask the experts. Justice Blackman might as well have asked a meteorologist when life begins, because that is how irrational his opinion is.

Justice Blackman goes on for a long paragraph talking about the various smart people who are not experts on when life begins. He mentions Stoics and Aristotle. He mentions Jews, Catholics, and Protestants. And when he considers the view that life begins at conception, instead of characterizing it as a view that comes from science, he wrongly characterizes it as a view that comes from Catholicism. In other words, Blackmun writes off the science as Catholic theology. So how does Blackmun avoid the science? He bizarrely ends the paragraph with a single sentence about embryology. In this sentence he simply states that conception is a process, not a singular point in time. For some reason he thinks this is a problem and a good enough reason for him to not consider the science and to write off the belief that life begins at conception as Catholic theology. It should cause some righteous anger in us that we have spent over forty years killing tiny humans, our offspring, the most vulnerable members of the human family, based on a gross dismissiveness of science.

There is no need to be confused about when life begins. We must simply look to the experts, biologists, to explain the science. Thanks to growth, development, reproduction with heredity, and the other characteristics of life, we can know with absolute certainty that life begins during conception. The question isn't and has never been about when life begins. The question has always been whether we will treat all humans with dignity and respect or whether we will divide humans into those who are equal members of the human family and those who are not.

Chapter Notes

1. Hochman, David. "The Playboy Interview With Cecile Richards" *Playboy* (2018)

   https://www.playboy.com/read/playboy-interview-cecile-richards

   Cecile Richards is the former president of Planned Parenthood Federation of America. (2006-2018)

2. Raven, Peter and George Johnson. *Biology, Fourth Edition.* Dubuque: Wm. C. Brown Publishers, 1996, pp. 1

3. Ibid, pp. 74

4. Schoenwolf, Gary C. , et al. *Larsen's Human Embryology*, 4th *edition.* Philadelphia: Churchill Livingston, 2009, pp. 5

5. Raven, pp. 73-74

6. Raven, pp. 73

7. Schoenwolf, pp. 43

8. Schoenwolf, pp. 44

9. Jacobs, Steven. *Balancing Abortion Rights and Fetal Rights: A Mixed Methods Mediation of the U.S. Abortion Debate.* 2019. The University of Chicago. PhD dissertation. pp. 237-256

   https://knowledge.uchicago.edu/record/1883

10. Lewis, Ricki. "When Does Life Begin? 17 Timepoints" PLOS *Blogs* (2013)

    https://blogs.plos.org/dnascience/2013/10/03/when-does-a-human-life-begins-17-timepoints/

11. Roe v. Wade. United States Supreme Court. (1973), pp. 160

    http://cdn.loc.gov/service/ll/usrep/usrep410/usrep410113/usrep410113.pdf

Fetal Beauty

HOW WE AGE PREGNANCY

Chapter 7

Estimating Gestational Age

One of the key concepts to understand when talking about embryo and fetal development is gestational age. When we talk about gestation we are talking about the process of growing inside the mother's womb. The fetus gestates or develops inside its mother. So when we talk about how far along a woman or fetus is in the pregnancy, we refer to that as gestation. For example, when we say that a baby is 26 weeks gestation, we are describing how long the baby has been growing in the uterus. We are also describing how far along the fetus is in its development.

When we refer to the gestation, we are not referring to the body of the fetus or embryo itself. This is an important distinction. The embryo is not a gestation. You can hold an embryo in your hand. You cannot hold a gestation in your hand. Gestation is something the embryo does. The embryo grows or gestates inside the uterus. The reason this distinction is so often confused is because the abortion industry so often purposefully and improperly uses the word gestation interchangeably with fetus or embryo. One such glaring example comes from the official abortion industry textbook, *Management of Unintended and Abnormal Pregnancy*, published by the National Abortion Federation, the official trade association of the abortion industry. The National Abortion Federation is to abortion what the National Rifle Association is to guns. The first sentence of chapter 6 states:

> *When a woman requests an abortion, she presumes she is pregnant with a viable intrauterine gestation, and she usually has some idea of its duration.*[1]

"Doctor, I think I'm pregnant with a gestation!" is probably something you've never said. Outside of the abortion industry, everyone refers to it as a fetus, embryo, or baby. All of those words are grammatically correct. Only in the abortion industry do they use words like gestation and pregnancy in a grammatically and medically inaccurate way so as to cause deliberate confusion and distraction. We will look more closely at the abortion industry's use of euphemisms like this one in chapters 9 and 17.

There are two standard methods of measuring the gestational age of the embryo/fetus: the method of time since the last menstrual period and the method of time since conception.

The method of time since the last menstrual period is the most common method and is sometimes referred to by its initials, LMP. The LMP method is calculated as the amount of time since the first day of the last menstrual period. Most women don't begin to suspect that they may be pregnant until their period is late. This means that most women are already at least four weeks pregnant with the LMP method before they even take a pregnancy test. The due date is calculated by simply adding forty weeks to the first day of the last period. This is the way most women estimate the gestational age of their pregnancies due to the fact that it is the easiest way to estimate. In order for this method to work, the woman has to have at least semi-regular periods and have some idea of when her last period started. But this is a very unreliable way to estimate age. According to the American College of Obstetricians and Gynecologists (ACOG), approximately half of women cannot accurately recall the first day of their last period and an early first trimester ultrasound is the best way to determine gestational age. The ACOG does not support estimating the age without an ultrasound, stating, "A pregnancy without an ultrasound examination that confirms or revises the [estimated due date before 22 weeks] of gestational age should be considered suboptimally dated."[2]

The second method of measuring gestational age is to estimate the time since conception. This method is typically used when studying embryology and fetal development as it more accurately describes development. It differs from the LMP method in that conception happens approximately two weeks after the last period. And so with the LMP method, for the first two weeks of pregnancy you aren't actually pregnant with an embryo yet. In fact, medical doctors don't actually consider you to be pregnant until conception or until the embryo has implanted in the uterus. This means that gestational age measured in LMP is about two weeks more than age after conception. Simply add or subtract two weeks to convert one method to the other. As you can see, estimating gestational age can become confusing very quickly.

It's about to get more complicated since conception does not take place exactly two weeks after the last period. It can vary by as much as five days in either direction depending on when the egg is ovulated and how long it takes to make its way into the fallopian tube and get fertilized. And so the LMP method is not a precise way to understand gestational age. You can't know precisely from your last period, how much your fetus or embryo has grown.

Estimating gestational age is a critical aspect of abortion. The age of the fetus determines what type of abortion and technic will be used as well as the price. For example, the RU486 abortion pill has been approved for use up to ten weeks gestation. I talked to one woman who went to a Planned Parenthood RU486 clinic. They estimated that she was at eleven weeks and told her she needed a surgical abortion. This lady decided that she would not go through with a surgical abortion and kept her baby. Vacuum aspiration abortions are also limited. If the fetus is too large, the body parts will not be vacuumed out successfully and a late-term abortion procedure must be performed.

Failure to estimate gestational age accurately can lead to significant complications. Dr. Warren Hern, one of the foremost experts on abortion in the United States, describes five major categories of abortion complications, one of which is "error in the estimate of length of gestation."[3] He goes on further to say, "An orderly review of the major sources of complications and their management must begin with a discussion of preoperative evaluation and accurate estimation of gestational length."[4] In other words, an important part of avoiding complications is to estimate the age accurately which, according to the ACOG, can only be done optimally with an ultrasound before 22 weeks gestation.

Unfortunately, not only are many abortionists not using ultrasounds to estimate age, some abortionists aren't even bothering to confirm pregnancy. I find it particularly shocking that a doctor would perform a surgical procedure on a woman who isn't even pregnant. Keep in mind that this information is not coming from a pro-life source! Dr. Hern is one of the most trusted experts on abortion in the country. He specializes in abortions in the second and third trimester and has been an abortionist since before Roe. He is a celebrated late-term abortionist in pro-choice circles. He says the following about abortionists failing to accurately diagnose pregnancy:

> *The first decision that must be made is whether the operation requires the patient to be pregnant. In most surgical procedures, at least a presumptive diagnosis is available. I believe the same standard should apply to abortion. However, this caution is regularly abandoned in "menstrual extraction." By claiming to "extract" the menses, practitioners make the diagnosis of pregnancy irrelevant and somehow justify the performance of abortion on non-pregnant women, as well as women in whom the pregnancy is too early to detect by hormonal assay. This serves both the patient and the physician badly for a variety of reasons.[5]*

> *A concomitant variation of this theme is the use of increasingly sensitive pregnancy tests. It is important to remember that highly sensitive*

tests will render a much larger proportion of false-positive results.[6]

Estimating gestational age is also critical in determining if the fetus is considered viable or not. Viability means that the fetus has a greater than fifty percent chance of survival if birthed. If it is viable, it may be illegal to do an abortion depending on the jurisdiction. If it is legal and viable, then the abortionist will give the fetus a lethal injection to ensure that the fetus is not accidentally birthed alive. According to ACOG, an ultrasound done around the time of viability, generally considered to be twenty-four weeks, can be off by as much as two weeks in either direction.[7] And so when a late-term abortionist does an ultrasound to determine if the child is viable, whether or not the child can be aborted, estimating age is not precise. It is really just a very educated guess. In other words, we determine when the woman's right to an elective abortion ends and when the child's right to live free of violence begins based on an educated guess that could be wrong up to two weeks in either direction. When you add sloppy and criminal abortionists like Kermit Gosnell to the mix, the results are shockingly horrific.

The estimated gestational age is also used to determine the price of the abortion. The cost of an abortion starts at about $350, depending on where you live, and can go as high as many thousands of dollars. Abortion is subject to the same market forces as other services. If there are fewer doctors doing abortions, then supply is down and the prices go up as a result. When fewer people want an abortion, demand goes down and so does the price. But when it comes to gestational age, the further along you are in the pregnancy, the more you will pay. This is because of the cost involved in doing abortions combined with the scarcity of abortionists willing to go further into pregnancy.

Let's look for example at an eleven week vacuum aspiration abortion versus a fourteen week abortion done with a sharp uterine curette or forceps. Vacuum aspiration is a relatively easy and quick procedure. You simply insert the vacuum curette and turn on the electric pump. The electric pump does most of the work. There can be complications, especially if the doctor is rushed and doesn't take proper precautions. But compared to the curette and forceps, it has a relatively low risk of complications. This is cheaper because you don't have as high of cost to manage the complications. This is especially true in terms of medical malpractice insurance. With a later abortion, there is more work involved. The abortionist has to go in and remove the fetus, usually in pieces, with his own strength. It is more work. It also has a significantly higher risk of complications. The sharp uterine curette and forceps are rigid metal instruments which carry more risk of damaging the uterus. The cervix has to be dilated more to make room for the instruments. These factors all contribute to the price.

The most expensive are the shockingly horrific third trimester abortions. There are only a handful of doctors in our country willing to do these. A very low supply means a higher price. The procedure takes longer. These babies have to be administered a lethal injection. The patient returns the next day to confirm the death of the child before proceeding to removing it. Some women will choose to be induced and deliver the dead baby. Others have the baby removed piece by piece with a forceps. Either way the time commitment from the doctor and staff is much higher. There is also the additional cost of the lethal drug used and the risk of complications is very high. Third trimester abortions can cost many thousands of dollars.

For the purposes of this book, unless I state otherwise, I am referring to gestational age in LMP, the number of weeks since the last menstrual period. And I am assuming that conception occurs exactly two weeks later, even though we know that conception can vary by as much as ten days. Keep that in mind as we look at the growth and development of the embryo and fetus in the next few chapters.

Chapter Notes

1. Paul, Maureen, et al. *Management of Unintended and Abnormal Pregnancy* Hoboken: Wiley-Blackwell, 2009, pp. 63

2. *Committee on Obstetric Practice, "Methods for Estimating the Due Date," American College of Obstetricians and Gynecologists* (2017)
    https://www.acog.org/Clinical-Guidance-and-Publications/Committee-Opinions/Committee-on-Obstetric-Practice/Methods-for-Estimating-the-Due-Date?IsMobileSet=false.

3. Hern, Warren. *Abortion Practice*, Philadelphia: J.B. Lippincott Company, 1984, pp. 175

4. Ibid, pp. 177

5. Ibid, pp. 177-178

6. Ibid, pp. 178

7. Committee on Obstetric Practice

THE MYSTERIOUS FIRST MONTH

Chapter 8

Embryonic Development in the First Four Weeks (Conception through Embryonic Folding) and Infertility Treatments

In chapter five we looked at the events leading up to and including conception. We ended with the chromosomes of the sperm and the egg fusing together and dividing into two new cells with a complete new set of twenty-three pairs of chromosomes in each cell. In chapters five and six we learned about reproduction with heredity and how conception makes this possible in the human species.

In this chapter we will explore the growth and development of the embryo through the mysterious first month. I call it the mysterious first month because these first four weeks after conception were the hardest for me to get my head around. We know that the embryo starts as just two cells after the DNA has fused together and divided. And we know that four to five weeks later the embryo has a three dimensional tiny body. But there is a lot of complex activity going on in the growth of that little embryo in between those two points. I hope I can help you comprehend this journey from two cells to a recognizable three dimensional body. The mysterious first month is an incredibly exciting time in the development of the tiny human child. Let's go briefly through this eye-popping journey together.

Now that the egg is fertilized, it is an embryo that makes a week long journey down the fallopian tube in search of the inviting lining of the uterus. At this point, the mother isn't considered pregnant yet by the American College of Obstetrics and Gynecologists (ACOG) even though she has already reproduced. Whether or not a woman is considered pregnant at fertilization or at implantation of the embryo into the lining of the uterus depends on which doctor you happen to ask. While the official position of the ACOG is that pregnancy begins at implantation, 57% of OBGYNs consider pregnancy to begin at fertilization. Only 28% consider pregnancy to begin at implantation. The remaining 15% were unsure, according to a 2011 survey published by the American Journal of Obstetrics and Gynecologists.[1]

As the little embryo makes its journey to the uterus, a number of things have to occur before it implants. During this journey, the embryo is safely protected by its shell. This is the same shell that allowed only one sperm to enter the egg while keeping the others out. As the embryo makes its way to the uterus, its cells continue to divide and multiply, causing it to develop further. Each time the cells divide, the result is double the number of cells. At this point there is an important concept to understand called cleavage. When the cells of the embryo divide and multiply it is called cleavage. What's important to understand about cleavage is that it does not increase the overall size of the embryo. As the cells divide, the new cells are smaller. So while there are more cells, the total size is the same. They continue to fit neatly inside the protection of the shell.[2]

Another important concept to understand is compaction. Compaction occurs in the last couple of days before implantation. Compaction is what forms the embryo from a bundle of distinct cells into a well-formed and more sophisticated body of cells. The cells form together to give the outside of the embryo a smoother surface. When looking at the embryo before conception, you will clearly see the separate and distinct cells which look like a bunch of round balls. After compaction they are formed together to have the appearance of only one larger round ball.[3]

Compaction also results in the cells changing into inner cells and outer cells. When a group of cells changes into different kinds of cells we call this differentiation. And so compaction results in one round cohesive ball of cells that are differentiated into inner cells and outer cells. This is important because the inner cells are what eventually will become recognizable as a baby.[4]

During this journey down the fallopian tube, the embryo also takes on fluid. This fluid collects in the center of the embryo. By the time the embryo makes its way to the uterus, it is a fluid-filled sac. In medicine, fluid-filled sacs are often called cysts. This is where the name blastocyst comes from. The embryo is now called a blastocyst. The embryo now consists of the shell, the outer cells which make up the sac containing the fluid, and the inner cells which are on the inside of the sac. You can imagine for a moment that the blastocyst is like a small beach ball. The outer cells are represented by the plastic walls of the ball. The air in the ball represents the fluid. Now let's imagine that you taped something like a quarter to the inside of the ball's plastic wall. That quarter would represent the inner cells.

The embryo has now completed its one week journey to the uterus. But it is not yet ready to implant into the uterus. It must first hatch out of its shell. The shell has done its job of only allowing one sperm cell inside the egg and protecting the embryo on this journey. But now its protection is no longer needed. The shell is now getting in the way of implantation. So the egg now makes a hole in the side of the shell and squeezes out. After hatching, the embryo is now ready to implant into that nice, inviting lining of the uterus. You can imagine for a moment that the embryo is like a seed and the lining is like freshly tilled and watered soil that is ready to sustain the seed. In order to implant, the embryo's outer cells divide and grow into the lining. The growth reminds me of the seed sprouting roots and the roots growing into the soil. The embryo grows into the lining. As it grows, it actually pulls itself under the surface of the lining and buries itself into the lining, reminding me of the way a gardener plants the seed under the surface of the soil.[5]

Once the embryo has started implanting into the lining, it begins producing a hormone called hCG. This hormone is what causes the mother to continue producing progesterone, which is needed to support the lining of the uterus and the developing embryo. The embryo produces increasing amounts of hCG for the first few weeks of pregnancy. This hormone is also what triggers a positive pregnancy test. The hCG is detected both in home pregnancy tests and blood tests. If a woman is trying to get pregnant, she may begin taking pregnancy tests each month to see if she is pregnant. The hCG levels may be high enough to trigger a pregnancy test about the time that her next period is due or soon after.

Now that we've looked at human development from fertilization through implantation, we can take a brief look at infertility and two common methods of treating it. Infertility is a common and increasing problem. But the treatments available are also increasing and improving at an astonishing rate, putting treatment within the reach of more people every year. One of the two common treatments is called intra-uterine insemination (IUI). This is simply artificial insemination. It is often the first treatment tried as it is minimally invasive, affordable, and relatively easy. IUI requires very close monitoring so that insemination will be done at the ideal time to get pregnant. For an IUI, a clinic staff person will process the sperm in order to get a concentrated collection of the best sperm. The sperm is deposited into the uterus. Essentially an IUI is taking the best sperm and shortening their journey by placing them in the uterus. The woman may also be given drugs to aid in ovulation. IUIs can increase the likelihood of sperm making it to the egg to fertilize it.

The second treatment is the one that most people recall when thinking about infertility, in vitro fertilization (IVF). It is a more difficult treatment in nearly every way. It is costly for most people. Even the cheapest options cost thousands of dollars and they can easily turn into tens of thousands of dollars. It is also stressful and emotionally difficult through each step of the process. How many eggs will fertilize? What if they don't implant? How are we going to pay for another round if it fails? All these questions and more weigh heavy on IVF patients. And then there are the risk of complications. IVF involves flooding the woman with hormones that can have all kinds of side effects such as headaches, mood swings, and vomiting. And finally there are the ethical conundrums to navigate. How much debt and stress are you willing to put the family through? How many eggs do you attempt to fertilize? What is your plan if you have leftover embryos? Will you attempt to implant them all, donate or adopt them out to another infertile couple, freeze them indefinitely, or destroy them? All of these questions and more must be considered. But the greatest questions involve how many new little lives you will create and your plan for them.

When a woman or couple go through IVF treatments, typically they join a community of people who are also going through IVF. That community might be friends who have done IVF or it might be a closed Facebook group where people talk about their experiences. In this community, you quickly learn the

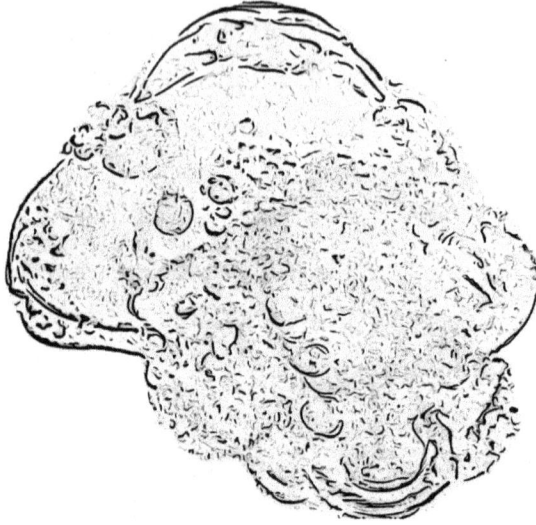

Our Daughter Lois Hatches From Her Shell

language developed just by this community to describe their experiences to each other. One such term is PUPO. This is an acronym for "pregnant until proven otherwise." Women use this to describe the period between the transfer of the embryo to her uterus and the pregnancy test. For this short time the woman knows that she has reproduced an embryonic child but does not know yet if it will implant into the lining of the uterus. For many women, the knowledge that her embryo is in her uterus is a deeply meaningful maternal experience even if it does not implant. Another term used in the IVF community is "embaby." This is the combination of two words, embryo and baby. The reason for this term is because the word baby carries along with it a sense of emotional attachment. When we talk about babies it provokes emotions in us and especially emotions of attachment between parent and child. The term embaby is a way for IVF couples to communicate their relationship and their emotional attachment to their child even as a tiny embryo. My wife and I were PUPO two times and our last embaby implanted successfully. Our earliest baby picture of our oldest daughter is a picture taken by the fertility clinic of her hatching out of her shell in preparation to implant in Felicia's womb. This wasn't a picture of random tissue or cells. It is a picture of our daughter. She was our daughter from the very beginning.

Now that we have looked at embryonic development in relation to fertility treatments, we are ready to look at implantation. During the fourth week, the embryo fully implants itself underneath the surface of the lining of the uterus. During this week the embryo is large enough to see with the naked eye and is only the size of a speck of dust. One of the key features of this week is the development of the beginning of the placenta. The structures that are needed to provide oxygen and nutrients to the embryo develop during this week. This allows the mother's blood to interact with the very beginning of the embryo's blood. The mother's blood gives oxygen and nutrients to the embryo's blood. The embryo returns waste to the mother's blood. The growth of this very early circulatory system while buried in the uterine lining is vital to the continued growth of the embryo. By the end of the fourth week, the embryo has grown and now looks a little larger than a speck of dust. The inner cells have now grown out a membrane, forming the amniotic sac. The inner cells that will become the baby, the embryo proper, are still in their place at the side of the embryo. At the end of the fourth week, the embryo proper is now large enough that it is visible as the size of a speck of dust.[6]

The key feature of the fifth week is the growth of the inner cells into the embryonic disc and the differentiation of the cells into various tissues and structures needed for the baby. At this point the embryo's body is in the shape of a flat oblong. This oblong shape is referred to as a disc. It is a flat or two-dimensional disc. It doesn't become a three-dimensional body till week six. From this point we can simply refer to it as the embryo. The embryo is growing in size, both in thickness and in width and depth of the disc. The cells differentiate meaning that they transform into different types of tissues and structures needed for the baby. Some examples include the heart, spinal cord, and brain. As cells differentiate into these various organs, they need to be able to move to their appropriate places in the disc so that they are ready to form the baby. The cells have fascinating hair like structures.

These tiny hairs spin in circles, causing the cells to move about and find their appropriate places.[7] Sometimes the cells don't migrate to the correct location. This is the case with a condition called Situs Inversus, when all the organs go to the reverse side of the body, creating a perfect mirror image of where they should be. Some famous people with Situs Inversus are Cathrine O'Hara, Enrique Iglesias, and Donny Osmond.[8] It's incredible to think of how your body intricately grew even as a tiny disc.

A key feature of the disc shaped embryo is a structure called the neural plate. The neural plate is the beginning of the nervous system of the embryo and will grow into the brain, spinal cord, and nervous. Even before the embryo becomes a three-dimensional body, the growth of the brain and nervous system has already begun. This neural plate runs on the upper surface of the disc-shaped embryo. It runs the length of the oval shaped disc but is not large enough to cover the whole disc.[9] At the end of the fifth week, the embryo is now over 1/16th of an inch long and growing rapidly.[10]

The embryo is now entering the sixth week of gestation and the final week of the mysterious first month. The various parts needed for the baby are in their places and ready to commence assembly. This stage of development is called folding. Folding is the growth of the disc into a three-dimensional body.

Folding is the perfect climax to the end of the mysterious first month as the disc folds up almost like magic into the body of something that could be described as starting to look like a tiny baby's body. It is the awe-inspiring moment when all the parts assembled in the disc rise in a choreographed display almost as if they were showing off what they can do. An analogy would be a flower bud blooming in spring. The various parts of the flower have already started developing in the bud. But at the appropriate time, the flower bursts forth in all its splendor. You could say that the embryonic disc is like a bud. And during the sixth week, the embryo bursts forth in the splendor of a tiny, incredible body.

The way that the disc folds up is a little difficult to imagine. You can envision the disc as a piece of paper. You fold the paper in the middle so that it is divided into two halves. You fold each half again so that it is divided into four. Now imagine that this paper is flat on the table and represents your disc. The center of each half rises up. The centers of each half fold up till they meet each other and the paper takes the shape of a tube. This is a simple explanation of embryonic folding which occurs in the sixth week of pregnancy.

By the time the sixth week is finished, the tiny embryo's body has some distinct visible features. The face is beginning to form, buds that are the beginnings of the arms and legs have formed, and the heart is a

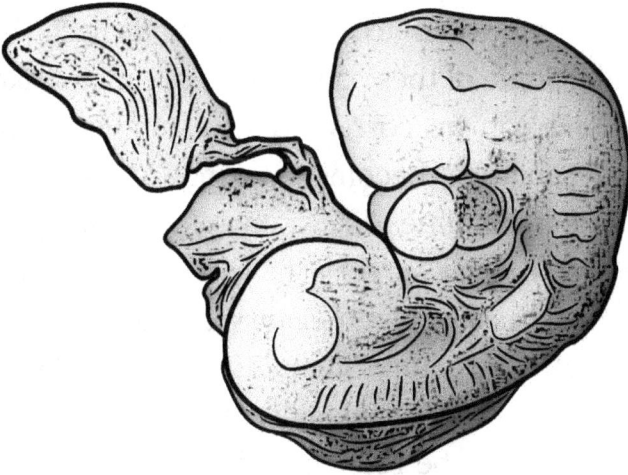

6 Week Embryo After Folding

dominant, visible feature. The heart has begun beating by the end of the sixth week and is easy to see due to the thin clear skin. The embryo is curled up at this point in a position that is known as the "fetal position" with the abdomen and heart in the center. The neural plate that was on top of the disc has now folded up into a neural tube running from front to back of the embryo. At the front it is the brain. Running back is the spinal cord. The embryo is now pushing a quarter inch long and the parts are visible to the naked eye.

Folding is a crucial point of development since many birth defects occur during this time. There are a wide variety of birth defects or disabilities that can result when folding does not complete properly. The most common is spina bifida. This is where a portion of the neural tube does not finish folding up in the back of the child. Some of these disabilities are minor like cleft palate and can be corrected with surgery. Some of these disabilities become lifelong disabilities. And sometimes disabilities, like spina bifida, can actually be treated with fetal surgery. In fact there is an entire chapter of *Larson's Human Embryology* that is dedicated to "the fetus as patient." Sometimes these disabilities are ultimately fatal. Shockingly, the risk of birth defects at this stage can be reduced by 75% simply by taking folic acid. Folic acid is a B vitamin and is taken as an individual supplement and as part of prenatal vitamins.[11] Folic acid is an essential nutrient needed for folding. If enough is not available, folding isn't completed properly. With the overwhelming importance of folic acid, it is disappointing that our society doesn't do more to educate women about and stress the need for folic acid. Many of these unborn children with birth defects go on to be aborted with elective abortions, even those with defects that are not fatal. We could do much to reduce late-term abortions and improve children's lives simply by increasing the use of folic acid.

Unfortunately, many women do not discover that they are pregnant and begin taking folic acid until it is too late. Most women do not discover that they are pregnant until four to eight weeks after their last period. At four weeks the embryonic disc is already forming and by eight weeks folding has since finished and the embryo has grown to look very clearly like a baby. In addition to the majority of women who discover that they are pregnant between four and eight weeks, a significant minority of women don't discover that they are pregnant until after eight weeks and even into the second or third trimester. This could be due to many reasons, including irregular periods, denial, lack of access to health care, lack of symptoms, and obesity. By the time a woman realizes that she is pregnant, folding may have already finished and the birth defect may already be present. This is why I urge women who are sexually active and of child bearing age to take a folic acid supplement even if they aren't planning to get pregnant. It's better to be safe and avoid the risk.

Over 300,000 abortions are done each year in the United States during this mysterious first month. According to the CDC, about 37% of abortions are done up to and including the sixth week.[12] Since few women discover that they are pregnant before the fourth week, it is safe to assume that nearly all this 37% are done in the fourth, fifth, and sixth week of pregnancy. This includes abortions during the fourth week when the embryo has its own blood and begins taking in nutrients from the mother's blood to its own blood. It includes abortions during the fifth week when the organs are developing in their anatomically correct places in the disc-shaped body. And it includes abortions in the sixth week when the body folds up into the fetal position, arm and leg buds form, and the heart begins its first beats. Many of these very early abortions do end a beating heart. Most of these early abortions in the United States are now done with the RU486 abortion pill which we will learn more about in chapter 13.

In this chapter, we have looked at the development of the first four weeks after conception ending with the three-dimensional body of the embryo with a face and a prominent beating heart. The various organ systems, tissues, and structures of the unborn child are in their places. All that is needed now is for them to grow and mature until the baby is ready to be birthed. We will be looking at this growth and maturation in chapter 10.

Chapter Notes:

1. Chung, Grace S. et al. "Obstetrician-gynecologists' beliefs about when pregnancy begins" *American Journal of Obstetrics and Gynecology*, 2012, Volume 206, Issue 2, pp. 132.e1-132.e7

    https://www.ajog.org/article/S0002-9378(11)02223-X/fulltext

2. Schoenwolf, Gary C. et al. *Larsen's Human Embryology*, 4th edition. Philadelphia: Churchill Livingston, 2009, pp. 41

3. Ibid, pp. 47

4. Ibid, pp. 43

5. Ibid, pp. 44

6. Ibid, pp. 62

7. Ibid, pp. 69

8. Solomon, Saskia. "Situs inversus and my 'through the looking glass' body" *The Guardian*. (2016)

    https://www.theguardian.com/science/blog/2016/sep/08/situs-inversus-and-my-through-the-looking-glass-body

9. Schoenwolf, pp. 95

10. Hill, Mark A. "Carnegie Stage 9" *University of New South Whales*. (2019)

    https://embryology.med.unsw.edu.au/embryology/index.php/Carnegie_stage_9

11. Schoenwolf, pp. 117

12. Jatlaoui, Tara C. "Abortion Surveillance-United States, 2015" *Center for Disease Control and Prevention*. (2018), Table 9

    https://www.cdc.gov/mmwr/volumes/67/ss/ss6713a1.htm#T9_down

Fetal Beauty

RECLAIMING THE WORD FETUS

Chapter 9

What a Fetus Is and Why it Matters

"Some products of conception may not fit through a smaller cannula."

Management of Unintended and Abnormal Pregnancy [1]

Many people, both pro-choice and pro-life, are surprised to learn that abortionists, like those in the quote above, avoid the words embryo or fetus. They especially avoid the word fetus. But to a certain extent they tend to avoid the word embryo as well.

One of the reasons this surprises people is that pro-choice advocates do often use the words fetus and embryo, believing that it helps their cause. Many on the pro-choice side, who have been conditioned to rely heavily on euphemisms, believe fetus or embryo to be useful euphemisms. You will frequently hear pro-choice people say things such as, "It's just a fetus, not a baby." They believe that these words can be used like so many other euphemisms to downplay the significance or importance of those tiny living humans. For a small but vocal segment of pro-choice people, fetus and embryo are used as derogatory terms. For this group of people, these words are used to denigrate or demonize tiny humans. For example, I've heard pro-choice people refer to embryos and fetuses as snot, trash, and parasites. For these people, the words "embryo" and "fetus" are used in the same way that a racist would call black people the "N-word" or a misogynist would call women the "B-word."

While some abortionists do readily use the words embryo and fetus, most are careful to avoid them. There is an aversion to these words that is very curious and peculiar. Further into this chapter, I'll explain what I believe is the motivation for this aversion. To see the great lengths they go through to avoid these words, one only needs to look at the textbook published by the National Abortion Federation, *Management of Unintended and Abnormal Pregnancy*. Chapter 10 of that textbook book describes in great clinical detail how to perform vacuum aspiration abortions. These are the suction abortions done in the first trimester and the most common surgical abortion in the United States. Vacuum abortions are done on both embryos and fetuses. You would think that this chapter must make some mention of the embryo and fetus. How can you write an entire chapter, going into great detail about how to do these abortions, without any mention of the fetus or embryo?

But the authors managed to do just that! They were meticulous in the way that they managed to avoid those two words. Throughout this chapter, they referred directly to the embryo/fetus over 50 times while carefully using techniques to avoid those two words. Some of those techniques include euphemisms like "tissue," referring only to specific parts of the embryo such as the "calvarium" (referring to the skull), and substituting an adjective for the noun such as the term "embryo-fetal parts." This last technique is the most telling. Abortionists are much more willing to use these words as adjectives than they are to use them as nouns. The authors used the adjective "fetal" four times and also used "embryo" as an adjective once. They don't want to confront the embryo or fetus directly. And so instead, they use these words as adjectives so as to give the appearance that the embryo or fetus isn't actually a thing. It's just a description. So the fetus instead becomes "fetal parts", "fetal cells", and "fetal tissue."

Some of the various words and phrases that they used in chapter 10 of this medical textbook to avoid the words embryo and fetus are as follows:[2]

- Uterine aspirate
- Fetal parts
- Intrauterine pregnancy
- Gestational sac
- Uterine contents
- Calvarium
- Tissue
- Earlier pregnancies
- Flow of tissue
- Fresh tissue aspirate
- The products
- Material
- Products of conception
- Pregnancy sacs
- Intact tissue
- Retained products
- Multifetal pregnancies (referring to more than one fetus, such as twins or triplets)
- Gestational tissue
- Major elements
- Very small embryo-fetal parts
- Persistent intrauterine pregnancy
- Persistent gestation

Now that we've established the great lengths through which abortionists are willing to go to avoid these two words, we must ask what motivates them. What is so bad about "embryo" and "fetus" that they must be avoided? I believe the answer is their humanity. Unlike pro-choice activists, these medical doctors are educated enough to know exactly what they are aborting. They know the biology well. They know that embryos are alive and are human. They also know that these embryonic humans are highly complex and sophisticated from the moment of conception. And they know that they are killing these little creatures. This is something they'd rather not talk about either amongst themselves or with their patients. And so an entire industry of abortion professionals meticulously avoids talking about it.

There are some abortion doctors that will use the word embryo, but not the word fetus. You could say that abortionists fall into three camps: those that use both embryo and fetus, those that only use embryo, and those that use neither. In the example above, the abortionists who wrote that textbook tried to use neither word. But what about the abortionists that only use embryo? What motivates them?

Near the end of the 10th week of pregnancy, medical doctors stop calling the prenatal child an embryo and start calling it a fetus. But why start at 10 weeks? Is there a specific type of development that happens at 10 weeks to trigger the change? No, there is not. If there is nothing specific at 10 weeks to justify calling it a fetus, then why do medical doctors and scientists pick this point? There is a general and widely recognized consensus to start calling it a fetus at 10 weeks despite the fact that there is nothing particularly special about this point of development. Because there is no objective point in development, professionals may vary by a few days as to when they call it a fetus. The reason they call it a fetus is because it looks like a fetus. In other words, you will know it when you see it. But what does it mean to look like a fetus?

If you take a look at a few pictures, it quickly becomes apparent that a fetus looks strikingly like a human baby. The eyes, the ears, the fingers, the toes, and the tiny beating heart all scream, "I'm a baby!" When looking at the fetus, you might have the same emotional responses that you would have for any other baby. It's cute. It's precious. It's so itsy bitsy!

Now the word baby isn't a scientific or medical term like embryo or fetus. The words fetus and embryo are technical medical terms taken from the Latin. As technical terms, they have specific technical and objective definitions. Baby, on the other hand, is not a technical word with a precise definition. It's definition and usage can vary from person to person. If you want to call the fetus a baby, then you can. If you don't want to call the fetus a baby, then you don't have to. Whether you call it a baby or not doesn't change what it is. Claiming that it is not a baby doesn't change its humanity. It doesn't cause it to look less like a baby. And it doesn't cause the fetus to be less endearing. The fetus is still the same human child regardless of whether you call it a baby or a parasite. We know this is true because people regularly refer to their fetal children as babies. People never show off ultrasound pictures of their wanted unborn children and call them fetuses. You will never hear someone looking at an ultrasound say, "Wow, your fetus has your nose!" When it's wanted, it is always called a baby.

The word fetus is at the intersection of humanity-affirming words used by pro-life people on the one side and the deceptive euphemisms used by pro-choice people on the other side. This has caused much confusion on both sides as to when and how to use this word in order to further their cause. On the pro-life side there are those who believe the word fetus to be yet another euphemism designed to dehumanize the child. And then there are people like myself who believe that we should claim medically accurate words like fetus and use them to educate the public about the humanity of the child. The pro-choice side is likewise divided. There are those who don't know what a fetus looks like and believe that they can use the word to downplay the child and justify aborting it. But there are others, the abortionists in particular, who understand that a fetus looks like a baby. These pro-choice people frequently avoid using the word fetus because it is an admission that they are killing humans that look like babies. The last thing an abortion doctor wants to do is to explain to his patient what her child actually looks like. If she realizes that it looks like a baby, she will be less likely to abort. And for the abortionist to call it a fetus is a reminder in his own mind that he is actually killing a baby.

It's no secret from the title of my book, *Fetal Beauty*, how I've decided to use this misunderstood word. I believe that pro-life people like myself should eagerly embrace technical terms like fetus. First, we should embrace it because it is medically accurate. Unlike euphemisms, this word actually has a specific and clear technical meaning. In other words, fetus is not a deceptive euphemism when used correctly. Second, we should aggressively educate people about the fetus and its humanity. Before researching for this book, I never knew that a fetus was called a fetus because it looks like a baby. Even in all my years growing up in a pro-life community, I was never taught this. I only learned it after reading a fetal development textbook. The pro-life community can and should claim the word fetus. We can explain to people that it is called a fetus because it looks and acts like a baby, whether the pro-choice side wants to admit that it is a baby or not.

Chapter Notes

1. Paul, Maureen, et al. *Management of Unintended and Abnormal Pregnancy* Hoboken: Wiley-Blackwell, 2009, pp. 147

2. Ibid, pp. 135-156

THE GROWING CHILD

Chapter 10

Fetal Development from Folding till Birth

In chapter 8 we looked at the development of the embryo from conception through the folding of the embryo into a three-dimensional body at six weeks gestation. In chapter 9 we looked at why we stop calling it an embryo and start calling it a fetus at ten weeks. Now that we know it is called a fetus because it looks distinctly like a little human baby, let's take a look at its development from embryonic folding till birth. At this point all of the various organ systems and tissues have started growing in their places. The rest of pregnancy is mostly spent growing the size of the fetus and maturing the various organ systems so that the fetus will be ready for birth.[1]

As you read through this chapter, it may be helpful for you to refer to the website for The Endowment for Human Development at www.EHD.org. This website is the most helpful I've found for understanding the complexities of early development. The best resource on this website is the collection of videos of real living prenatal humans. These videos virtually take you inside the womb, face to face, with real unborn children. On this website, you can also dig deeper into the things I discuss in this chapter. This organization also has a fantastic app for your smartphone or tablet called See Baby. The See Baby app is also a great resource for pregnant mothers who want to follow their babies' growth through pregnancy.

One of the more exciting and recent technological developments in embryology is the use of digital 3D imaging to map early human embryos, allowing us to see the vast intricate organ systems in the tiny embryo. It is called "Tridimensional visualization and analysis of early human development." Researchers published their images in the journal *Cell* in March of 2017. The images show the development of unborn children in striking detail starting at six weeks gestation. You can view the images and videos at www.transparent-human-embryo.com.

What this technology allows researchers to do is take digital pictures of very tiny features of the embryos. It then distinguishes different cell types and digitally colors the cells that the researcher wants to highlight. It can even isolate cell types so as to create images of only that part of the fetus. They do this extensively with the nerves. For example, they took an image of a finger with the nerves in the finger isolated and colored. Then they erased the finger so that you only see the nerves. Researchers were able to create a "map" of the tiny embryonic and fetal bodies. Some of the striking images that they published in the journal *Cell* include nerves, muscles, fingers, toes, brains, facial features, taste buds, and lungs. The images that are most awe inspiring are the taste buds and the nerves. Most people are surprised to discover a well-developed tongue and taste buds at nine and a half weeks gestation. Most people are even more surprised to see the development of the nerves, spinal cord, and brain this early as well. The nervous system begins growing before the embryo even folds up into a three-dimensional body. The earliest embryo that they mapped was at six weeks gestation. At this point there is already a vast system of nerves throughout the entire body. Thanks to these images, you can now see that vast network of nerves growing and branching out even further until eleven weeks gestation.[2] From six to eleven weeks gestation, when these images were captured, is the time in development when most abortions occur.

We will start our look at the growth of the prenatal child with the skin and the skin's related parts: hair, teeth, and glands. After the embryo folds up, its skin has only a single layer of cells. At ten weeks we begin calling the embryo a fetus and it is beginning the fetal stage of development. At the fetal stage, the skin is paper thin and translucent. It is now only three layers of cells thick. But it continues to add layers and thickness throughout pregnancy. In the jaw there is already the beginning of twenty temporary "baby teeth" and twenty permanent adult teeth. They start as little buds and don't finish growing until after birth. Throughout the rest of pregnancy, the teeth grow larger and the enamel develops. The adult teeth continue to grow as little buds beneath and at the side of the temporary baby teeth. The adult teeth are already tiny buds soon after folding. At the beginning of the fetal stage, the first hair follicles form and most of the hair follicles, approximately five million, are present by about twenty two weeks. The buds that are the mammary glands, or breasts, grow throughout the fetal stage. The buds grow larger in size until about twenty-six weeks when in both sexes the milk ducts develop. The finger nails begin as thickenings on the fingers at ten weeks. The toe nails begin at fourteen weeks.[3]

Bone and muscle growth begin as soon as folding has occurred and by the fetal stage the bones and muscles are well underway. Most but not all of the bones begin as cartilage. Cartilage grows throughout the rest of pregnancy, turning into bone. The primary job of bones and muscles is to grow larger in size until they are ready for birth. As the skeletal system supporting the body grows, the baby is able to grow larger as well. The skeletal system finishes growing when adulthood is reached.[4] During pregnancy it is important that the mother provides lots of calcium for the developing child's bones, especially in the third trimester when so much growth is happening. At around eighteen to twenty weeks, the child will have grown big enough that the mother can feel him kicking and moving inside her womb. The child's bones are softer than those of adults and its skull is made of several bones that haven't grown together yet. This is to allow for the difficult journey through the birth canal.[5]

The brain, spinal cord, and nerves are all in their places and growing at the beginning of the fetal stage. They began as the neural plate on top of the embryo before folding. Folding causes the neural plate to fold up into the shape of a tube. The front of the tube is the beginning of the brain with the rest of the tube being the beginning of the spinal cord. The rest of pregnancy features rapid growth of the gray and white matter of the brain. The gray matter is on the outside and is what we are used to seeing in pictures of the brain. The white matter is just underneath the gray matter. These grow out of the cerebellum, a part of the brain in the lower back near the spinal cord, and grow to surround the rest of the brain. They divide into four lobes initially but continue to grow into many smaller lobes within the four lobes to create the wrinkled look that we all have come to recognize from pictures of the brain. The very thin outermost layer of gray matter is called the cerebral cortex. This is where rational decision-making occurs. The brain, and especially the cerebral cortex, don't finish maturing until adulthood.[6] Remember this the next time your teenager wants to make a bizarre decision, like announcing that she is moving to Africa to save baby lions. Yes, the lion saving example was inspired by a true story!

Interestingly, there is good evidence that fetuses have brain activity and experiences in the womb that are similar to that of newborn babies, particularly after seven months gestation.[7] By seven months, unborn babies appear to be dreaming. They sleep most of the time and alternate between rapid eye movement (REM) sleep and non-REM sleep.[8] REM sleep suggests that they are dreaming and scientists speculate that their dreams consist of sensations they have experienced up till that point in the womb. Fetuses may actually dream earlier in pregnancy, but the evidence doesn't exist yet to know. We also know that they begin hearing sounds as early as sixteen weeks.[9] And they can actually

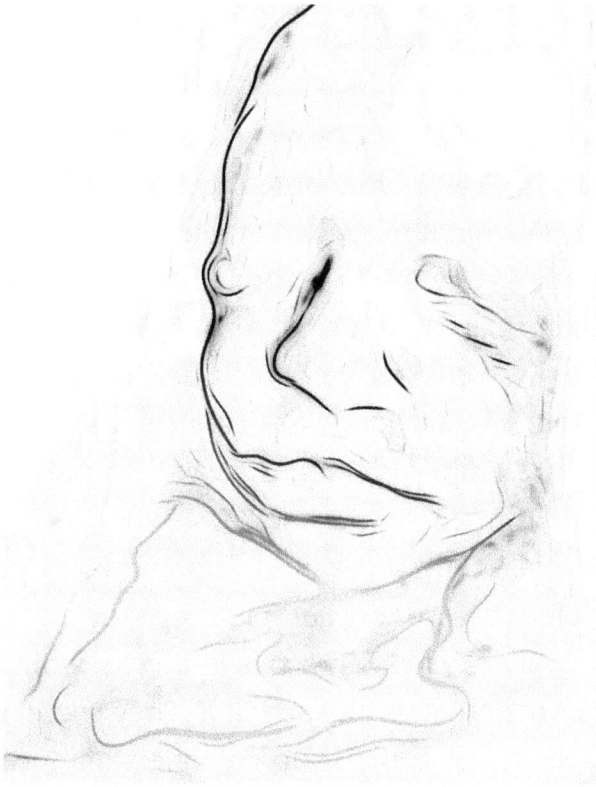

3D Ultrasound of our Daughter Lois at 26 Weeks

recognize language and music. These unborn babies actually react to the sounds they hear and recognize familiar sounds like their mothers' voices or familiar songs.[10] Not only can they recognize sounds, they appear to recognize experiences in the womb. In one fascinating story, twins at twenty weeks appeared to be playing with each other while divided by a membrane. A year after birth, they mimicked this experience by playing with each other on either side of a curtain. It appeared that they were recreating this experience in the womb.[11]

We also know that brain activity in the unborn child begins much earlier as an embryo thanks to the work of researchers in the 1950s. The fact that embryos have electrical activity in the brain is not new information. We knew this over a decade before Roe v. Wade. Two researchers were able to study several of these living embryos after they had to be surgically removed from their mothers. The youngest embryo was just over eight weeks gestation and was removed due to a life threatening ectopic pregnancy. They were able to apply electrodes to this embryo's brain and do an EEG to detect brain activity. They were able to detect electrical activity up to ninety minutes after being removed from their mothers. Not only did they detect electrical activity but they detected different patterns in the firing of the neurons. This means that the neurons did not appear to be firing randomly but instead in an orderly manner. The earliest embryo who they tested was only about two weeks after folding.[12]

While this embryo was just over eight weeks gestation, brain activity likely starts a little earlier. Good evidence of this is that we can see the embryo moving under its own power on an ultrasound as early as seven weeks. If the embryo is moving then that

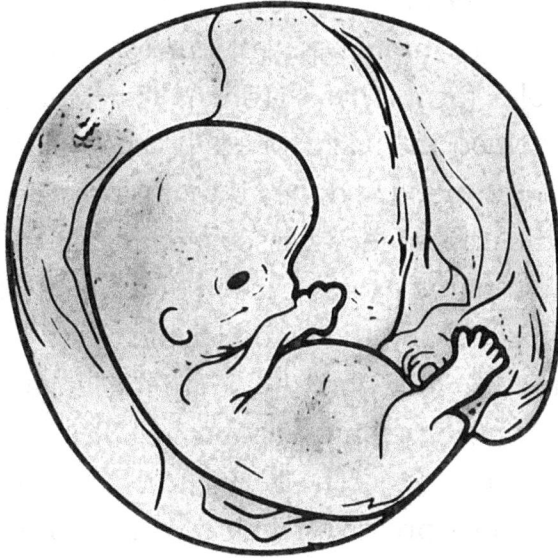

8 Week Embryo in Its Gestational Sac

means that electrical activity must be causing those tiny muscles to move. These movements begin first as arching of the body. But by the end of the first trimester, their movements become quite sophisticated and include hiccupping, stretching, and yawning.[13]

Another important organ to develop is the lungs. The lungs are typically the last to develop of the vital organs needed for the fetus to be born and survive after birth. The lungs begin as a single tiny bud soon after folding occurs. The bud branches out as it grows. By the time it reaches the fetal stage, the lungs are already branching out like the branches of a tree. There are already two distinct lungs. At about sixteen weeks, the lungs enter a new stage of development. At the very ends of these branches, structures called respitory bronchioles form. These bronchioles come in contact with the blood and make it possible to exchange gasses between the blood and the air that is breathed into the lungs. The exchange of gasses means that the blood is able to discharge excess carbon dioxide into the lungs to be expelled, and the blood is able to take in oxygen from the air that was breathed into the lungs. The stage of growth of the respiratory bronchioles is sixteen to twenty eight weeks. The very earliest that babies are able to be born, breath on their own, and survive occurs around the middle to end of this stage of lung development.[14]

At this stage, the fetus begins producing a substance called surfactant. Surfactant makes it possible for gasses to pass through the thin walls of the respiratory bronchioles. If a doctor suspects that a baby will be born prematurely, he may give the mother steroids in order to speed up growth of lung tissue and production of surfactant.[15] Felicia and I have personal experience using steroids for our youngest child. Both of our biological children were born at thirty-seven weeks. The older child had to stay in the NICU for assistance in breathing. For the younger child we asked for steroids. The steroids successfully helped her to avoid being admitted into the NICU.

In addition to steroids, doctors may also give the premature babies surfactant replacement therapy. This simply means that additional surfactant is put into the baby's lungs allowing gasses to exchange.[16]

In a healthy fetus, the ability to breathe and exchange gasses is the final step in development needed to make it possible for the baby to survive after birth and the cutting of the umbilical cord. Before the umbilical cord is cut, the child exchanges gasses with the mother in the placenta. After the cord is cut, she must be able to exchange gasses in the lungs in order to survive. When a doctor estimates that the fetus is likely to be mature enough to exchange gasses in the lungs and survive, that is called viability. A baby isn't considered viable until it has a greater than 50% chance of survival. If a baby has a 30% chance of survival, it isn't considered viable. Further, it's impossible for doctors to know for certain if a fetus is mature enough. At best they make an estimate. What this means is that babies that are not considered viable can sometimes survive after birth. It also means that viability is a moving target. Thanks to modern medicine, viability is generally around twenty-four weeks. But without the help of modern medicine, viability would be more like thirty-six weeks. And so viability is a moving and vague standard.

This is very important as the Supreme Court has decided that elective abortion is a constitutional right up until viability. The Supreme Court has said that states can prohibit abortion after twenty-four weeks on the basis that viability is generally around that time. Essentially what the Supreme Court did is to say that the mother's right to abort ends and the human rights of the child begin when the child is able to exchange gasses in the lungs. In other words, your ability to exchange gasses is what actually gives you your human rights. And because lab created surfactant makes it possible to exchange gases earlier in pregnancy, it is actually the surfactant that gives you your human rights. So in practicality what the Supreme Court did in Roe v. Wade and Planned Parenthood v. Casey was to declare that our human rights come from a chemical created in a laboratory. When you break the idea down to its essence, the viability line in the sand that the court drew is indefensibly arbitrary and vague.

So what would happen if modern medicine developed to the point that babies could survive earlier? It would likely mean that abortions could be prohibited earlier assuming that the Supreme Court sticks to its viability claim. The idea that we could hook a very tiny baby in the second trimester to a machine to replace the placenta sounds like sci-fi. But it isn't a far-fetched idea. Consider dialysis, for example. We routinely replace people's kidneys with a machine that processes blood so that people can continue to live. Is it really a stretch to think that we could hook a very tiny baby up to a machine in order to exchange gasses without either the placenta or the lungs? I believe we will be seeing this kind of technology in the future.

The fetus prepares for life outside the womb by practice breathing. It practices by breathing or swallowing amniotic fluid. The practice breathing can begin as early as ten weeks, the beginning of the fetal stage of development, and continues through birth. This practice breathing is very important to the development of the lungs. When you have an ultrasound of your unborn child, you may be able to see it practice breathing.[17]

The heart begins to beat by six weeks gestation. This isn't a matter of opinion. It is a scientific fact. *Larson's Human Embryology* says that the "heart begins to beat" at 22 days after conception.[18] And yet the pro-choice movement and especially main stream pro-choice media outlets are refusing to admit this simple scientific fact. Why is this so controversial? It has to do with the recent trend by pro-life states to pass laws prohibiting abortion after a heartbeat can be detected. These "heartbeat bills" have been highly effective in persuading the public that abortion should be

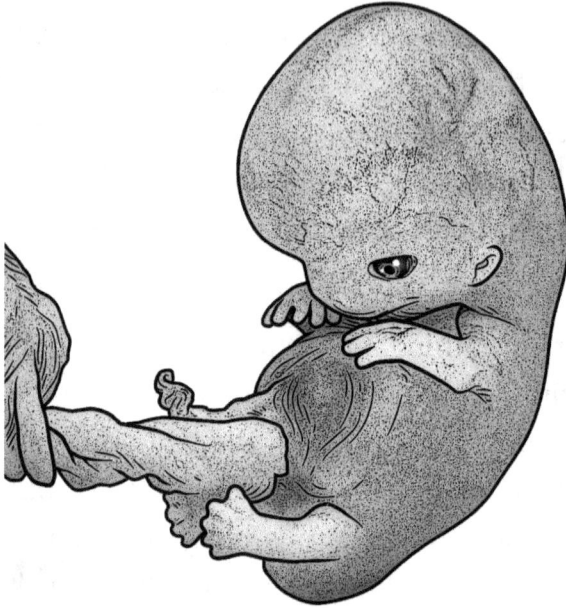

8 Week Embryo with Beating Heart

prohibited earlier in pregnancy. And so an all-out effort is being waged by the pro-choice movement to deny that a heartbeat is a heartbeat. Instead of calling them heartbeat bills, they insist that we call them "fetal pole cardiac activity" bills.[19]

When the embryo folds up, the early heart tissue in the disc folds up into a tube shape. Very soon after folding the tiny heart begins beating. Even before the embryo becomes a fetus at ten weeks it has blood pumping through veins and arteries. The heart has a prominent place on the front of the abdomen and can be easily seen by the average observer. By the time the heart reaches the fetal stage of development, it has looped and developed valves looking much more like the shape we think of when we think of the heart.[20]

Now let's take a look at the face. About a week after folding, the face of the embryo begins to appear visibly recognizable. The beginnings of the nostrils, eyes, and ears can be seen on the face. The tongue has also begun to develop. Below the nostrils are several swells or arches that are turning into various parts of the face and head. At this point the face may look odd to anyone who is not used to looking at embryos. But by the time the embryo reaches the fetal stage at ten weeks, the face has developed dramatically so that it looks clearly like a baby. The eyes are large and dark in color. They now have lens and eyelids. At the beginning of the fetal stage the eyelids begin sealing shut and won't open again till the third trimester. The ears, nose, and mouth look more distinctly like a baby. The tongue now has taste buds that are visible through a microscope. And the fetus both hiccups and begins practice breathing.[21]

By eleven weeks the face has become quite active. The eyelids have finished fusing together. The fetus can now sigh, suck its thumb, and move its jaw and tongue. By twelve weeks there is eye movement, yawning and vocal chords have begun to develop. By thirteen weeks he can now smile. He is starting to have complex facial expressions that will continue to become more elaborate throughout pregnancy.[22] The fetus has been known to do some odd and inexplicable behaviors, one of which is licking the wall of the uterus.[23]

The reproductive system of the fetus has developed, according to its sex, by about ten to eleven weeks gestation. At this point the various reproductive parts are in place including the penis, testes, uterus, and ovaries. At this point in pregnancy, with an advanced ultrasound machine and a skilled technician, you may be able to see the sex of your baby. In the girl fetus, her germ cells have already begun dividing to form mature eggs for reproduction. These cells won't finish creating mature eggs, however, until after puberty.[24]

The most exciting part of embryo and fetal development may be the limbs. By the fetal stage at ten weeks all the parts have developed: arms, legs, hands, feet, fingers, and toes. The limbs begin first as tiny buds right after folding occurs at six weeks. Two leg and two arm buds first appear on the side of the embryo. These buds then lengthen into arms and legs with pads on the ends for hands and feet. Next the pads develop fingers and toes in them. At this point nerves have already grown into the fingers and toes connecting them to the brain. The fingers and toes are now webbed. We are now at eight weeks gestation and the embryo can move its arms, legs, hands, and feet. Movement is an important part of the growth and development of the limbs. Soon after, the webbing dies away, leaving just the fingers and toes. Next the fingers and toes grow longer and more slender, looking more like our fingers. By week twelve fingernails and unique fingerprints have developed.

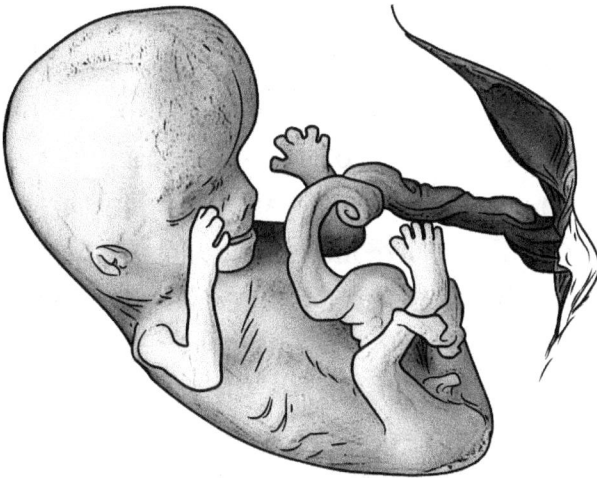

Fetus at the End of the First Trimester

The latest scientific research in embryology has found an exciting new discovery, that embryos develop many muscles in their hands and fingers that you and I no longer have. Researchers created and studied images of the hands between seven and thirteen weeks gestation. They were able to highlight the muscles in the hands to create stunning images. They found thirty muscles at seven weeks. By thirteen weeks, the number dwindles to only twenty muscles. Some of these muscles fuse together to form a single muscle, while others simply disappear. These muscles give the embryo more dexterity in its fingers much like we have in our thumbs. If these muscles were to remain, we would also have more dexterity in our fingers. In this one aspect of human development, the human embryo is actually more sophisticated than you and I.[25]

The most contentious aspect of fetal development is fetal pain. At what point does the fetus actually feel pain? There are two reasons this is so contentious. The first reason is because fetal pain is one of the primary reasons cited by pro-lifer advocates to pass laws prohibiting abortion after twenty weeks. The second reason is because pain is only partly physiological or material. There is no way to objectively measure pain. I can't put a thermometer in your mouth or a cuff on your arm to measure pain.

In one respect, it doesn't matter if the fetus can feel pain. Even without pain, an abortion is still a violent act. If I were to take away your ability to feel pain, that wouldn't give me the right to kill you. But in other respects, it does matter. Pain is an additional injustice on top of the injustice of killing. If I were to torture you first, that would be worse than if I just killed you. And so the question of fetal pain is relevant to abortion and to fetal surgery.

When you go to the doctor and you are in pain, what does the doctor ask you to do? She asks you to point to a chart on the wall with frowning faces. The larger the frown, the more pain you are in. You are asked to point to the face that best describes how much pain you have. Why must you do this? The reason is because pain is partly in your body and partly in your invisible being. The body does give some evidence of pain. But the body can't actually tell us very much about a patient's experience of pain. You can't measure a patient's pain with a medical device like a thermometer or blood pressure cuff. There is no objective way to measure pain. Instead, the patient must form a subjective opinion about the amount of pain he is experiencing and then communicate that subjective opinion to the doctor.

This becomes much more difficult when the patient can't communicate. What about a newborn or an infant who is only a few months old? They can't communicate and tell us how much pain they are experiencing. So how then do we know that they are experiencing pain? We know because their bodies show evidence of pain. You could say that their bodies communicate non-verbally. The obvious example is a baby crying. But infants may show other signs as well. The baby's heart rate may increase. She may try to get away from the source of pain. Any of a number of bodily reactions to pain could be non-verbal cues to tell us that she is in pain.

The same is true not just of babies after birth but also of babies in utero. The body of the fetus can react to pain in different ways, giving us non-verbal cues that she is likely experiencing pain. Three of these reactions include facial expressions, stress hormones, and body movements. Facial expressions simply mean that the fetus's face looks like it is in pain as a result of painful stimuli. The most vivid example is the ability of the fetus to mimic crying in utero as early as twenty weeks gestation. Researchers were able in 2005 to actually capture and publish ultrasound video of a fetus crying in utero. I would encourage anyone who doubts that a fetus can feel pain to watch that video of a crying baby in the third trimester.[26]

Stress hormones are hormones that the body produces when in pain. I call these fight or flight hormones because they prepare the body to evade danger. For example, if you are walking down a dark alley and are startled by a sound, your body would release hormones to prepare you to either fight back or flee. Your heart beat would increase to make sure that your vital organs have enough blood flow. Adrenaline causes you to have more energy. Researchers have found similar stress hormones in the fetus as early as eighteen weeks. They also found that blood flow was

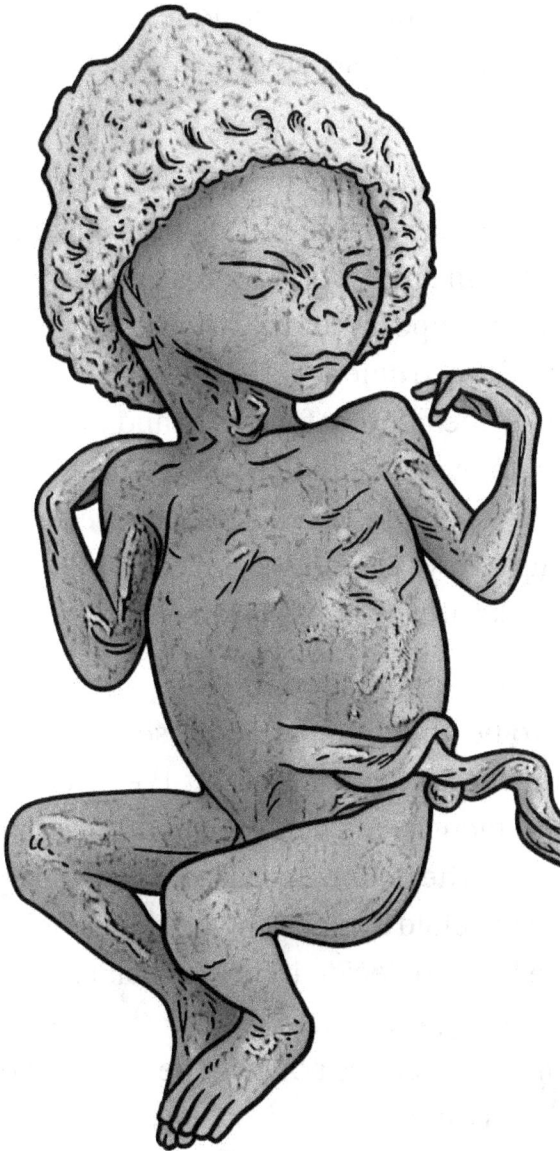

19 Week Fetus After Birth

prioritized to the fetus's vital organs.[27, 28] I've found the presence of fight or flight hormones at eighteen weeks to be particularly persuasive.

The final non-verbal cue is movement in response to painful stimuli. This may be the most powerful evidence that has caused abortion clinic staff to leave the industry. In an abortion a fetus or embryo may jerk, struggle, or appear to try to get away from the abortionist's instruments. The possibility that a fetus may actually be fighting for its life and that it may have some will to live is incredibly powerful. The fetus first begins to respond to touch at eleven weeks gestation. For example, if you lightly touch the sole of its foot at eleven weeks, it may bend its knee and curl its toes.[29]

While the argument over fetal pain on the surface appears to be an argument about science, the crux of the argument isn't actually about the science. Instead it is an argument over whether or not the fetus deserves the benefit of the doubt. Because we can't be absolutely sure what the child is experiencing in his or her invisible being, we have to decide if the possibility or likelihood of pain is enough reason to protect the fetus. The pro-choice movement does not give the child the benefit of the doubt. In their opinion, just because the fetus could feel horrific pain is not enough reason to spare it. The pro-life movement, on the other hand, does believe the fetus deserves the benefit of the doubt. In their opinion, the burden of proof is on the abortionist to prove that the fetus can't feel pain. You and I would certainly agree that we deserve the right to the benefit of the doubt. If you and I deserve the right to be treated humanely, why shouldn't the fetus be treated with equal protection? As you can see, this isn't so much an argument about science as it is an argument about equality for the fetus.

Now that we have finished looking at the development of the fetus till birth, the next several chapters will look at different abortion procedures. Understanding the development of the fetus is foundational to understanding how the various abortion procedures are performed.

Chapter Notes

1. Schoenwolf, Gary C. et al. *Larsen's Human Embryology*, 4th edition. Philadelphia: Churchill Livingston, 2009, pp. 167

2. Belle, Morgane. et al. "Tridimensional Visualization and Analysis of Early Human Development" *Cell.* (2017)

    https://doi.org/10.1016/j.cell.2017.03.008

3. Schoenwolf, pp. 193-209

4. Schoenwolf, pp. 217-244

5. Murkoff, Heidi. "Fetal Development: Baby's Bones and Skeletal System" *What to Expect.* (2018)

    https://www.whattoexpect.com/pregnancy/fetal-development/fetal-bones-skeletal-system/Schoenwolf, pp. 247-291

6. Hopson, Janet. "Fetal Psychology" *Psychology Today.* (1998)

    https://www.psychologytoday.com/us/articles/199809/fetal-psychology

7. American Institute of Physics. "Baby's First Dreams: Sleep Cycles Of The Fetus." *ScienceDaily* (2009)

    https://www.sciencedaily.com/releases/2009/04/090413185734.htm

8. López-Teijón, Marisa. et al. "Fetal facial expression in response to intravaginal music emission" *Sage Journals.* (2015)

    https://doi.org/10.1177/1742271X15609367

9. Skwarecki, Beth. "Babies Learn to Recognize Words in the Womb" *Science.* (2013)

    https://www.sciencemag.org/news/2013/08/babies-learn-recognize-words-womb

10. Bilich, Karin. "Baby's Alertness in the Womb" *Parents.*

    https://www.parents.com/pregnancy/stages/fetal-development/babys-alertness-in-the-womb/

11. Furth, Katrina. "Fetal EEGs: Signals from the Dawn of Life" *Charlotte Lozier Institute.* (2018)

    https://lozierinstitute.org/fetal-eegs-signals-from-the-dawn-of-life/

12. "When Does the Fetus's Brain Begin to Work?" *Zero to Three.*

    https://www.zerotothree.org/resources/1375-when-does-the-fetus-s-brain-begin-to-work

13. Schoenwolf, pp. 322

14. Schoenwolf, pp. 326

15. Schoenwolf, pp. 326

16. Watson, Kathryn. "How Do Babies Breathe in the Womb?" *Healthline.* (2017)

    https://www.healthline.com/health/pregnancy/how-babies-breathe-in-the-womb#fetal-breathing-practice

17. Schoenwolf, pp. 338

18. Schoenwolf, pp. 338

19. Gunter, Jennifer. "Dear Press, Stop Calling Them 'Heartbeat' Bills And Call Them 'Fetal Pole Cardiac Activity' Bills" *HuffPost.* (2016)

    https://www.huffpost.com/entry/dear-press-stop-calling-them-heartbeat-bills-and-call-them-fetal-pole-cardiac-activity-bills_b_584f19fde4b04c8e2bb14e15

20. "Prenatal Form and Function – The Making of an Earth Suit" *The Endowment For Human Development.* (2001-2019), Unit 8

    https://www.ehd.org/dev_article_unit8.php

21. *The Endowment For Human Development,* Unit 11

    https://www.ehd.org/dev_article_unit11.php

22. Bilich

23. *The Endowment For Human Development,* Unit 9

    https://www.ehd.org/dev_article_unit9.php

24. *The Endowment For Human Development*, Unit 10

    https://www.ehd.org/dev_article_unit10.php

25. Lanese, Nicoletta. "Tiny 'Lizard-Like' Muscles Found in Developing Embryos Vanish Before Birth" *Live Science*. (2019)

    https://www.livescience.com/disappearing-muscles-human-embryo.html

26. Gingras, J. et al. "Fetal homologue of infant crying" *Arch Dis Child Fetal Neonatal Ed.* (2005)

    https://www.ncbi.nlm.nih.gov/pmc/articles/PMC1721928/

    Follow the link to download the video of the fetus crying.

27. Smith, R. et al. "Pain and stress in the human fetus." *European Journal of Obstetrics & Gynecology and Reproductive Biology.* (2000)

    https://www.ncbi.nlm.nih.gov/pubmed/10986451

28. Blickstein, Isaac & Inbar Oppenheimer. "Compassionate Treatment of Fetal Pain" *Allied Academies.* (2016)

    https://pdfs.semanticscholar.org/c281/fdc4afbe255245dc086e519cd1fbace7b398.pdf?_ga=2.181623207.1479769947.1573469325-1181950952.1566205080

29. *The Endowment For Human Development*, Unit 9

    https://www.ehd.org/dev_article_unit9.php

Part C

ABORTION

THE CURETTE

Chapter 11

Curettage Abortions, Uterine Perforation, Internal Bleeding & Trauma Experienced by Abortion Workers

---

It is fascinating to watch the development to maturity of a new human being by some technique such as sonar to which I have devoted much of my research life.

The observed miracle of healthy development is enhanced by recognising the cases in which it goes wrong, often recurrently so. It is particularly heart breaking to witness the present day wanton policy in many centres of discarding into the bucket or incinerator so much healthy unborn life whose only fault is that it is unwanted.[1]

Dr. Ian Donald

When I picked up the curette, the first thing I noticed was the weight. This instrument had some unexpected heft to it. The next observation I made was of the sensation of the cold steel. There is something about that sensation that reminds you of the seriousness of this instrument. The one I picked up came as a set of three different sizes. Each one had a long handle designed for gripping. The handle had depressions along it, each one allowing its user to get a firmer grip. Beyond the handle was a rod which extended the scraping edge well beyond the handle so as to be able to use the scraping edge inside the uterus. Finally, I ran the scraping edge along my finger. To call it a scraper isn't entirely accurate. It's actually a blade that is a tad bit sharper than a butter knife. It's not sharp enough to cut my skin, but it is sharp enough to do some serious scraping. The blade reminds me of a teardrop extending from the rod. It is a rounded blade with both sides ending into the rod so that there are no sharp points. A sharp point inside the uterus could be disastrous.

The instrument I've just described for you is the Sharp Uterine Curette. The action of using a curette is called curettage.[2] A great variety of curettes are used throughout healthcare. What they all have in common is that they are all used to remove material by scraping. The curettes that most people recognize are the dental curettes used by dental hygienists to clean teeth. The hygienist will have an assortment of scrapers in different shapes that she uses to scrape the buildup of plaque off of teeth. The dental curette is different from the uterine curette in that the uterine curette has a distinctive blade for scraping.

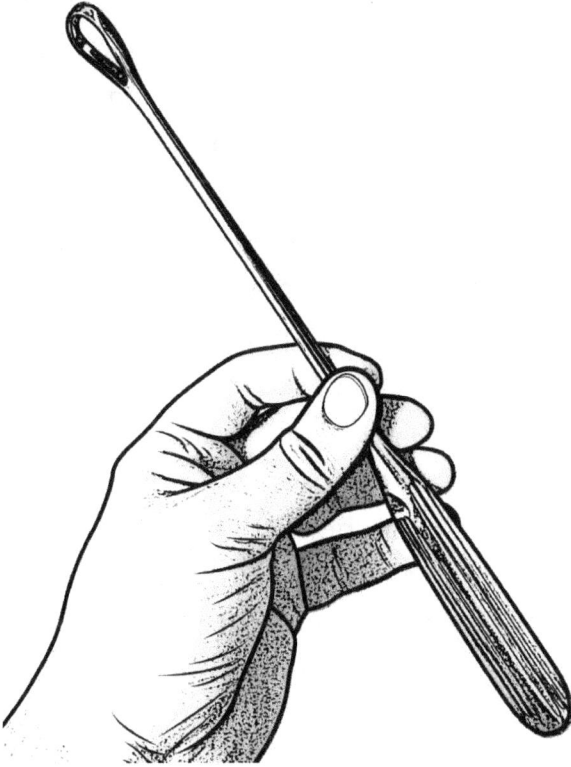

Sharp Uterine Currette

Before the introduction of the vacuum aspirator for abortions in the mid-twentieth century, the curette was the primary instrument used to bring about the death of the child and remove it. When I refer to the "primary instrument" in an abortion, I am referring to the instrument that is actually responsible for the death of the embryo or fetus. In order to understand what an abortion is, we have to begin with the curette.

When the curette is the primary instrument used, abortion doctors call it a "D&C" procedure. The term D&C refers to dilation of the cervix and curettage. The problem with this term is that it is so intentionally vague that you don't even know whether or not it is referring to an abortion. The D&C procedure is done for many legitimate health reasons other than abortion. Often it is used on women who are not pregnant to remove the lining of the uterus for unrelated health reasons. Further, a D&C could also be used to remove a fetus that miscarried. This is not an abortion because the D&C isn't causing the death of a fetus. The fetus has already died. The abortion industry engages heavily in vague terms like D&C because it conveniently distracts the patients, doctors, and staff away from the subject of the D&C abortion, the death of the unborn child.

A curettage abortion begins the way most surgical abortions do. The patient and the room have to be prepared for surgery. The cervix is numbed and manually dilated to allow the instruments to be inserted into the uterus. The preparation varies depending on the type of surgical abortion. I won't spend much time in this book on preparation as this isn't the controversial part of abortion.

With a curette abortion, the basic idea is simple. The curette is used to scrape throughout the uterus. In some cases the embryo may be scraped out intact. If

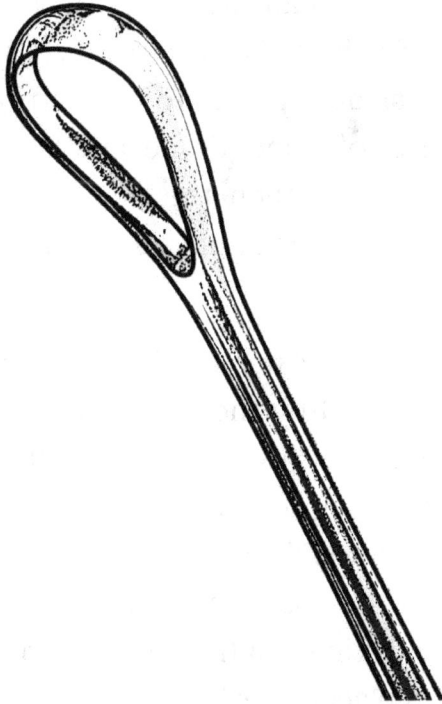

The Currette's Blade

this is the case, the embryo will die as a result of being separated from the life giving lining of the uterus and may survive for some time outside the uterus. The embryo or fetus is dependent on its mother through the placenta. The placenta allows oxygen and nutrients to transfer from the mother to the child. When the child is separated from the mother intact, it loses its source of oxygen and nutrients and dies relatively soon after. Everything else is scraped out as well, including the uterine lining, the placenta, and the umbilical cord. In most cases, the embryo or fetus does not come out intact. The scraping can cause the fetus to be scraped out in pieces. The blade is used to separate the body of the fetus into smaller pieces to be scraped out of the uterus in a mass of blood, tissue, and fetal body parts. When this happens, then the death of the fetus is caused by the massive physical trauma of being dismembered.

The trauma may not be limited to the physical trauma experienced by the fetus. There may also be psychological trauma experienced by the abortionist and staff. Depending on how large and developed the fetus is, they will visibly see the body of the child that was aborted. If the abortion is early enough, you may not be able to distinguish the embryo from the bloody lining, sac, and placenta as it is scraped out. But a fetus in the latter half of the first trimester can be clearly seen as it is removed. As the blade is removed from the uterus, you may see a dismembered arm or leg wrapped around it. If you are especially unlucky, you may even see a beating heart or tiny face looking up at you as it is removed. In surgical abortions, it is very hard for the doctor and staff to avoid seeing the destruction they have caused to the unborn child's body.

The abortionist or staff has to take steps to ensure that everything has been removed. If something like an arm or leg is left in the uterus, it will become infected and can become deadly in a matter of days. To make sure everything is removed, the abortionist may hold everything up against a backlight so he can make out the various body parts and confirm that they are removed. In order to lessen the stigma of this violent act, the doctor or staff may number the body parts or use other tricks to avoid using the real names. So instead of saying, "We are missing the head", the staff member may say "We are missing number 1." This trick and many others like it are common in the abortion industry to dehumanize the child.

If the abortionist or his staff confront the fact that they are killing tiny humans, they may experience regret and quit doing abortions. An abortionist by the name of Dr. Anthony Levantino is a classic example of this. He appeared in a video on the website *www.abortionprocedures.com* saying:

> *In the early part of my career I performed over twelve hundred abortions. One day, after completing one of those abortions, I looked at the remains of a preborn child whose life I had ended, and all I could see was someone's son or daughter. I came to realize that killing a baby at any stage of pregnancy for any reason is wrong. I want you to know today, no matter where you're at or what you've done, you can change.* [3]

Another such example is that of Dr. Paul Jarrett who did abortions in the year following Roe v Wade. He had the following to say about his last abortion in which he had to switch from using a vacuum curette to a forceps due to the size of the fetus:

Initially, the abortion proceeded normally. The water broke, but then nothing more would come out. When I withdrew the [vacuum] curette, I saw that it was plugged up with the leg of the baby which had been torn off. I then changed techniques and used ring forceps to dismember the 13 or 14 week size baby. Inside the remains of the rib cage I found a tiny, beating heart. I was finally able to remove the head and looked squarely into the face of a human being - a human being that I had just killed. I turned to the scrub nurse standing next to me and said, "I'm sorry." I knew then that abortion was wrong and I couldn't be a part of it any longer. No one was critical of me for what I had done, nor for having stopped. But I had a lot of guilt about that abortion and had flashbacks to it from time to time. I sometimes dreamed about it. The guilt lasted about four years. [4]

Notice how Dr. Jarrett talks about flashbacks and dreams about the abortion. Simply feeling guilty doesn't explain flashbacks and nightmares. But psychological trauma certainly could explain the flashbacks and nightmares. They could even be symptoms of post-traumatic stress resulting from the trauma of looking into the face of the aborted child. These kinds of testimonies resembling post-traumatic stress are not uncommon among doctors and staff who have left the abortion industry.

I can draw from my own experience to help explain post-traumatic stress and how it causes flashbacks. At one time I worked at a company where my job included operating a forklift. One day there was a tragic accident while I was operating the forklift. The result was a massive injury to a person's leg. The incident emotionally crippled me, but not because I felt any guilt. I knew it was an accident and that I wasn't to blame. As soon as first responders arrived I headed straight to the manager's office to try to regain my composure. One of my coworkers tried to assure me that it wasn't my fault. I responded that I knew it wasn't my fault, but it didn't make it any easier. For many days afterward I experienced flashbacks of the accident. At first I experienced them constantly. They tortured me. I relived the accident over and over again in head. I couldn't will myself to stop experiencing them.

Eventually they became less frequent until I stopped having them altogether. The reason I tell this story is to explain why I'd expect abortion workers to experience flashbacks. I only witnessed one mangled leg. Many of these workers have to witness all the mangled body parts: arms, legs, torso, and even the tortured faces. I don't see how anyone could not suffer traumatic stress from doing abortions.

We've looked at the physical trauma to the fetus done by the curette and the psychological trauma to the abortionist and staff. But we haven't looked yet at the psychological trauma to the mothers. We will look at this trauma further in chapter 12 regarding the abortion pill RU486.

There are a number of complications which can result from an abortion. Two common complications which can result in the death of the mother are uterine perforation and incomplete abortion. Uterine perforation is when the abortionist creates a hole in the uterus with an instrument like the curette. The abortionist can damage nearby organs. Uterine perforation can cause hemorrhaging or massive blood loss. In the most severe cases, the blood loss results in the death of the mother. Incomplete abortion is when parts from the prenatal human or her supporting structures are left in the uterus after an abortion.

These parts can cause infection. In the most severe cases, the infection turns into sepsis, organ systems shut down, and the mother dies. In this chapter we will look at uterine perforation. We will look at incomplete abortion and sepsis in the next chapter.

Uterine perforation is when an instrument inserted into the uterus causes a hole in the wall of the uterus. All surgical abortions carry a risk of uterine perforation because all surgical abortions involve instruments in the uterus. It's important to understand that the uterus is a relatively fragile organ. You could think of the uterus like a stretchy bag. As the fetus grows, the bag stretches out to make room. When instruments like the curette are used inside the uterus, these instruments cause the walls of the uterus to stretch. What this means is that inserting instruments into the uterus has inherent risks. Perforations are relatively common in surgical abortions and are usually very minor and heal on their own. Many are so minor that they go undiagnosed. But this isn't always the case. Sometimes perforations are much more significant and won't heal on their own. If the perforation is too large, the patient may experience a large amount of blood loss. In the worst cases, the instrument may go through the wall of the uterus and damage surrounding organs such as the bladder or colon.

Very few abortion clinics are capable of treating a woman with a badly perforated uterus. Typically these women must be rushed to the emergency room for surgery. It is not uncommon for abortion clinics to call 911 and send their patients to the emergency room in an ambulance. At some of the least safe clinics, calling for an ambulance becomes a regular occurrence. A Planned Parenthood abortion clinic in St. Louis Missouri has become known for its regular emergencies, calling for an ambulance 74 times over ten years. It averaged an emergency every seven weeks.[4] If a woman with a significant perforation is not treated quickly, blood loss and infection could become very dangerous and even result in death. She could die from blood loss or from infection.

Dr. Paul Jarrett described a patient who came to him with a massive perforation that ultimately led to her death. This 18 year old young lady had flown to the State of New York for a legal elective first trimester abortion. At this time New York was one of only a few states that allowed elective abortions before Roe v. Wade. One of the arguments the pro-choice side frequently makes is that legalizing abortion will make it safer for the women. I can't emphasize enough that this was a legal abortion and that she was one of many women who lost their lives to legal abortions in the years leading up to Roe v Wade. He described the young woman's situation saying:

When she returned home in terrible pain, she realized she was in trouble and for the first time, told her mother what had happened to her. Her mother contacted her own gynecologist, who in turn referred the patient to Coleman Hospital to be evaluated by the resident on call-me. Even though I was still wet behind the ears, I know that this pale, frightened little girl was still 10 weeks pregnant and her blood count was only half of what it should be. The private, attending doctor came in and took the patient to surgery immediately that night, where he repaired the hole that had been torn in the back of her uterus, which had caused her massive internal hemorrhage. Over the course of the next few days, infection set in which did not respond to antibiotics, and we made the painful decision to perform a hysterectomy. Tragically, the shock from the infection severely damaged her lungs and her course was steadily downhill. As I helplessly watched, she slipped into unconsciousness and a few days later she died.[6]

This story is of a woman who died from infection after her uterus was perforated. In some cases, women die from massive internal bleeding after a uterine perforation. That is what happened to a young woman in my home state of Delaware. Gracealynn Harris was a nineteen year old African-American woman who already had a young son. She went to abortionist Dr. Mohammad Imran for a second trimester abortion at his abortion clinic in 1997. After the abortion, she was weak and left the clinic in a wheelchair. Before the end of the day she was dead from massive internal bleeding. No one understood how badly hurt she was until that evening when she had a seizure. The family sued Dr. Imran and his clinic and were awarded 2.2 million dollars. Dr. Imran was found to have committed malpractice in part because he did not use an ultrasound to guide his instruments. An ultrasound guided abortion is one where the ultrasound machine is used so the abortionist can see what he is doing with his instruments. Most abortions are not guided. You could call these blind abortions. Gracealynn Harris could still be alive today if her abortion had been ultrasound guided according to an expert witness.[7]

It is important to understand that not all abortions are equally risky. There are a number of factors that go into the quantity of risk that a woman is exposing herself to with an abortion. But unfortunately, the abortion industry isn't being honest with women and the public about the risks. Instead of allowing women to decide for themselves the amount of risk that they want to take, the abortion industry loudly proclaims that abortion is safe. And they have an army of journalists, politicians, and entertainers to push this propaganda for them. There is a certain amount of condescendence in this claim that abortion is safe. That's because safe is a subjective opinion based on the amount of risk you are comfortable taking. And so when the abortion industry says that abortion is safe, what they are really saying is that the risk of harming you is a risk the abortionist is willing to take. But what about the amount of risk that you are willing to take? Abortion isn't about the woman and her doctor making an informed decision together. This is an entire industry pushing their lack of risk aversion on America's women.

There are a number of red flags that women can look for to indicate that they are taking unnecessary risks. Here are some factors to look for in order to reduce the risk of complications.

Is the clinic very clean? If you see blood on the floor, insects, or anything dirty like that, you are looking at an unsafe clinic.

Does the doctor take time to build a patient doctor relationship? If the doctor doesn't take time to get to know you, your concerns, and your medical history, he may be an unsafe abortionist.

Are the staff and doctor rushed? Unfortunately, many clinics are having staff prepare patients for their abortions and you only see the doctor for ten minutes when he comes in to do the abortion. A rushed doctor is an unsafe doctor.

Does the doctor do an ultrasound before the procedure? Doing an ultrasound beforehand is an incredibly important part of avoiding complications. What if you have twins? What if you have an ectopic pregnancy? There are so many things that can go wrong if they don't do a proper examination with an ultrasound.

The same is true of ultrasound guided and blind abortions. Does the doctor use an ultrasound to guide his instruments? Very few doctors do ultrasound guided abortions. The abortion industry considers ultrasound guided abortions to be optional and up to the personal preference of the abortionist. But if you want to avoid risk, you should demand it. Ultrasound guided abortions are far less risky than blind abortions.

How far along are you into your pregnancy? Abortions get much riskier as the fetus gets larger. This is especially true after the first trimester when the fetus is too large to be vacuumed out.

What is the doctor's record? Has he been disciplined by your state or local authorities? Has he lost any of his patients? What kinds of malpractice lawsuits has he faced and what were the results?

Unfortunately, most women will not be asking these questions before they get an abortion. But as you can see, the amount of risk a woman takes with an abortion depends on a lot of factors. To make a blanket claim that all abortions are safe is irresponsible and a disservice to women. Women should be informed on a case by case basis the amount of risk involved and offered options like ultrasound guidance of instruments to reduce risk.

Finally, the abortion is not intended to be safe for the fetus or embryo. In this chapter we've looked at a specific type of abortion using a sharp uterine curette. The purpose of this instrument is the remove the embryo or fetus, usually in pieces. The blade of the curette is used to dismember the fetus and remove the parts: head, torso, and limbs. When other instruments are used to kill the fetus, the curette is often used to clean up and make sure nothing is left inside the uterus. Once you understand the curette and how it is used in abortion, you are now ready to learn about other types of abortion.

Chapter Notes

1. Willocks, James and Wallace Barr. *Ian Donald: A Memoir* London: RCOG Press, 2004. pp. 120

   https://books.google.com/books/about/Ian_Donald. html?id=6GtmNEpf2fwC

   Dr. Ian Donald (1910-1987) was an accomplished Scottish OBGYN, pro-life advocate, and inventor of the fetal diagnostic ultrasound.

2. Curette is pronounced [kyoo-ret].

   https://www.dictionary.com/browse/curette?s=t

   Curettage is pronounced [kyoo r-i-tahzh].

   https://www.dictionary.com/browse/curettage?s=t

3. "Aspiration" *Abortion Procedures: What You Need to Know*. Live Action

   https://www.abortionprocedures.com/aspiration/

4. Jarrett, Paul. "Testimony of Dr. Paul Jarrett, Former Abortion Provider" *Priests For Life*.

   http://www.priestsforlife.org/testimonies/1125-testimony-of-dr-paul-jarrett-former-abortion-provider

5. Downs, Rebecca. "Planned Parenthood: Women Need An Abortion Center That Can't Keep Them Safe" *The Federalist*. (2019)

   https://thefederalist.com/2019/06/03/planned-parenthood-insists-women-need-abortion-facility-ridden-medical-emergencies/

6. Jarrett

7. Vidmar, Neil and Valerie Hans. *American Juries* Amherst: Prometheus Books, 2007. pp. 83-87 https://books.google.com/books?id=jrhCTNdVgSAC&lpg=PA1&dq=american%20juries&pg=PA1#v=onepage&q=american%20juries&f=false

THE VACUUM

Chapter 12

Vacuum Aspiration Abortions, Incomplete Abortion, Infection, & Sepsis

Now that we've looked at the curette and how it is used in abortion, we can look at the most common surgical abortion today: vacuum aspiration. Vacuum aspiration is any abortion where the fetus/embryo, as well as everything else in the uterus, is sucked or vacuumed out. While vacuum aspiration continues to be the most common surgical abortion, the abortion pill is increasingly being used instead.

The primary instrument used in this abortion is a combination of two instruments: the curette, which is an edge for scraping, and a suction tube, which is called a cannula. It is often called a vacuum curette or a vacurette. You are already familiar with the sharp uterine curette from the previous chapter with its metal blade for scraping. In the vacuum curette, the scraping edge of the curette is incorporated into the suction tube.

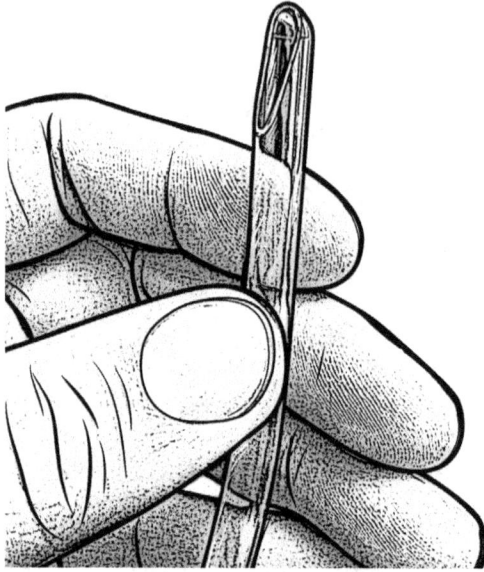

Typical Disposible Vaccum Currette

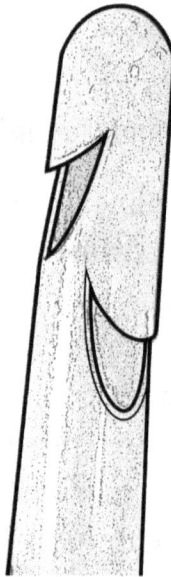

Double-Sided Vaccum Currette

The vacuum curette comes in many varieties. It comes in metal or plastic, straight or curved, rigid or semi-rigid, and reusable or disposable. It also comes in different sizes which are measured in millimeters. The style of vacuum curette used is left to the personal preference of the abortionist.

Vacuum aspiration can be further subdivided into aspiration with an electric pump versus manual vacuum aspiration which uses a hand-held syringe. Most vacuum abortions in the United States are done with the electric pump. In this abortion, the woman is prepared and the cervix is dilated and numbed as is necessary in surgical abortions. The doctor inserts the vacuum curette into the uterus and positions it. Then he turns on the electric pump in the aspirator. The vacuum curette is connected to the pump with plastic tubing. The electric pump is approximately ten times stronger than a household vacuum. It is very powerful and noisy. While the electric pump is turned on and creating a vacuum, the vacuum curette isn't engaged yet. The final step to begin the vacuuming is for the abortionist to engage a valve. Switching the valve applies the vacuum against the tubing and vacuum curette.

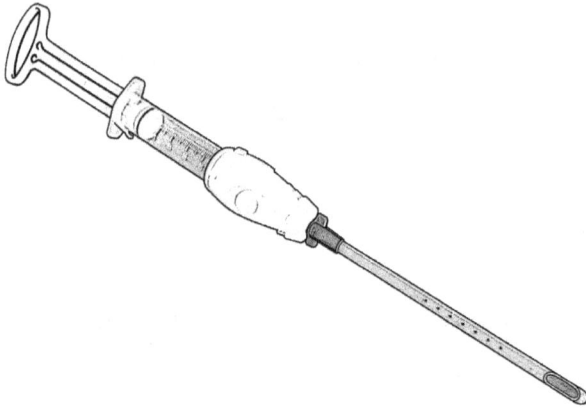

Manual Vacuum Aspirator

Manual vacuum aspiration uses a large hand held syringe. It is often referred to by its abbreviation, MVA. The vacuum curette is attached directly to the end of the syringe. MVA is nearly silent and the force of the suction isn't as strong. The type of vacuum used, electric or manual, is often left to the personal preference of the abortionist. While the electric pump is still the tool of choice for most abortionists, MVA appears to be growing in popularity both among abortionists and their patients.

Most abortionists use the electric pump because of the convenience and because they believe it to be less risky. The suction is more powerful which makes their job easier. In contrast, MVA requires significant arm strength to operate the syringe. MVA doesn't provide as much suction strength which means that

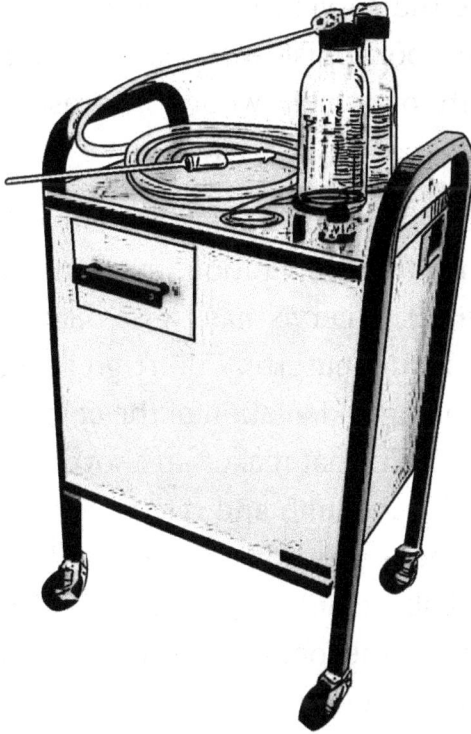

Electric Vacuum Aspirator

the doctor might have to use the syringe several times to get everything out. On the other hand, some doctors do prefer MVA over the electric pump because it is mostly quiet. The sound of the electric pump can be disturbing for the mothers, especially if the mother understands what the pump is doing to her fetus. MVA also allows the abortionist to have more direct control over the suction. Abortionists disagree over whether or not MVA is more likely to cause complications. It may require the cervix to be dilated more and it may require the abortionist to spend more time in the uterus, both of which would increase the risk of complications.

A vacuum abortion begins like any other surgical abortion with the preparation, numbing, and dilation of the cervix. The cervix has to be dilated enough to insert the vacuum curette. I don't go into great detail into the numbing and dilation of the cervix in this book because that isn't what makes an abortion an abortion. If all you did was numb and dilate the cervix, no one would care about abortion. There would be no pro-life movement if all we did was operate on the cervix. The embryo/fetus is the focus of the abortion. The abortion industry, on the other hand, focuses heavily on the cervix. They see it as a convenient distraction away from the embryo/fetus. That's why they name their procedures names like dilation & curettage (D&C) and dilation & evacuation (D&E). The goal is to keep you focused on the dilation of the cervix and away from the violent end of your embryo.

After dilation, a typical vacuum aspiration abortion begins with the abortionist inserting the vacuum curette into the uterus. The abortionist positions the curette where he thinks it should be. In the rare instance that an ultrasound is used to guide the abortion, the curette is placed exactly where it needs to be to get the embryo/fetus. The tubing from the electric vacuum aspirator then slides over the other end of the vacuum curette. The tubing goes back to the aspirator which has two glass jars where the fetal remains are collected. The aspirator is then turned on. The pump makes a creepy sort of humming and thumping sound. It's particularly disturbing because we know what it is about to do. If a woman suffers from post-traumatic stress from her abortion, the memory of the sound of the aspirator may be attached to her trauma. Similar sounds like the sound of a household vacuum turning on may startle her or cause flashbacks.

Now the pump is turned on but the vacuuming hasn't begun yet. When the abortionist is ready, he has to engage a valve on the aspirator that directs the suction against the tubing and curette. The bright red blood from the lining of the uterus can be seen through the tubing and into the jars. If he damages the uterus, the blood from the bleeding uterus is vacuumed as well. In most instances an ultrasound is not used to guide the abortionist. He moves the vacuum curette back and forth in a twisting motion to blindly scrape the inside of the uterus. If it is ultrasound guided, he can move the vacuum curette to get the fetus without blindly scraping.

If he fails to make contact with the embryo/fetus, the abortion is called a failed abortion because the fetus continues to live and grow. The woman is still pregnant. When the abortionist does make contact, the force of the suction removes the embryo in pieces. The arms, legs, torso, and head are all suctioned out piece by piece through the tubing and into the jars. Also suctioned out are the supporting structures: the umbilical cord, placenta, and membranes. Sometimes the head presents a problem for the abortionist. It is the hardest part to remove due to being large and round. If the child is too large, the head will not be suctioned out. The abortionist has to use a forceps and sharp uterine curette to crush it and remove it. Further, if the fetus is too large, other body parts such as a leg or arm may get stuck in the end of the vacuum curette and force the abortionist to switch to a forceps abortion.

The contents of the jars are then taken to be examined. Usually this is done in a separate room designed specifically for the purpose of examining the remains. Some clinics call this the "POC room," which stands for products of conception. POC is just one of the many ways abortionists and their staff dehumanize the child so as not to face the reality of what or who they just aborted. A more accurate name would be something like the fetus examination room. At this point someone must examine the remains to make sure that everything was removed. The abortionist or staff member holds the remains against a back light in order to see everything. Did they get the head? Did they get two legs and two arms? Did they get the rib cage? Staff members may also invent dehumanizing code language to describe the parts. For example, an arm might be described as #2. So if they only have one arm, they may tell the abortionist that they are missing a #2. The amount of destruction to the fetal remains differs between electric and manual vacuums. The electric vacuum is stronger. The body is likely to be dismembered into a larger number of pieces. Some of those pieces may be destroyed beyond recognition. MVA, on the other hand, is more likely to result in an intact embryo or fetus being removed. The weaker suction is less destructive to the body.

When an abortionist misses a fetal part or supporting structure and leaves it in the uterus, it is called an incomplete abortion. In an incomplete abortion, the fetus is successfully killed. This is in contrast to a failed abortion where the fetus survives and continues to grow. Incomplete abortion is one of the more common abortion complications, along with uterine perforation and hemorrhaging.[1] Incomplete abortion is made worse by the fact that most abortionists do the abortion blind, not ultrasound guided. When done blind, the abortionist cannot see if he got everything until it is examined against the backlight. But even with the backlight, it isn't hard to miss something. It all comes out as a mass of tissue and blood. While the untrained eye can identify an arm or a leg, it takes a trained person to identify everything and make sure that it is all removed. According to *Management of Unintended and Abnormal Pregnancy*, "Failed attempted abortion is usually recognized by immediate gross tissue examination with backlighting or the use of magnification when necessary. Clinic staff members can be trained to become proficient examiners of tissue specimens."[2]

An incomplete abortion has the potential to become dangerous very quickly. This is especially true if the uterus has become injured in addition to the incomplete abortion. The abortionist's instruments can injure the uterus. But the left over pieces from an incomplete abortion can injure the uterus as well. For example, a sharp bone fragment left in the uterus can cut into the side of the uterus. An incomplete abortion, especially when coupled with uterine injury, can become infected. A women with an infection from abortion needs to be treated right away. If she is not treated, the infection can turn into sepsis.³ Sepsis is when the body is extreme in its response to infection. The body starts a chain reaction that can result in tissue damage and entire organ systems shutting down. Ultimately, sepsis will result in death.⁴ That is why it is so important that abortion patients quickly get treatment if they suspect infection or sepsis. What started as a routine abortion can turn deadly in only a few days if left untreated.

One of the more well-known cases of a woman dying from sepsis from an abortion is the case of Marla Cardamone. She was an 18 year old from Allegeheny County Pennsylvania who went to the prestigious Magee Women's Hospital on August 15, 1989. She was there under pressure from a social worker for a late-term abortion. The social worker had erroneously convinced her that her baby would have birth defects from the medication she was taking. She received a urea installation abortion which involves killing the fetus with urea before inducing labor to deliver the dead baby. This type of procedure is explained in chapter 16. Unfortunately for Marla, the labor did not go as planned. Instead of delivering her dead child, she became badly infected. The infection became sepsis. As her body was shutting down, she suffered from violent seizures and vomiting which wrecked her body.

In only a matter of hours she was dead. Marla's mother had the following to say about her daughter's death.

> *Finally, they allowed me to see Marla's body. When I entered the room, I could hardly believe what I saw. There was my beautiful daughter so horribly disfigured that she was almost unrecognizable. A tube was still protruding from her mouth and I could see that her teeth and gums were covered with blood. Her eyes were half opened and the whites of her eyes were a dark yellow. Her face was swollen and discolored a deep purple. The left side of her face looked like she had suffered a stroke. All I wanted was to hold her. I managed to get an arm around her and kissed her good-bye.*

Marla's story didn't end like so many other women who have died from legal abortions. That's because her family decided to do something bold. They released the autopsy photographs to the public. The pictures were chilling. Her body was bruised and swollen. Her teeth and mouth were caked with blood. But the most disturbing picture was the one where the person doing the autopsy opened her uterus and showed her dead child still inside her. As if to add insult to injury, the hospital did not release the child's body to the family for a proper burial. Instead it was disposed of as medical waste.[5] The autopsy photographs can be seen at www.safeandlegal.com.

So far we have looked at curette and vacuum abortions as well as the emotional burden on abortion staff and the risk of incomplete abortion, infection, and sepsis. In the next chapter we will be looking at the abortion pill and the emotional trauma experienced by abortion patients, especially from the abortion pill.

Chapter Notes:

1. Paul, Maureen, et al. *Management of Unintended and Abnormal Pregnancy* Hoboken: Wiley-Blackwell, 2009, pp. 228

2. Ibid, pp. 227

3. Ibid, pg. 228

4. "What is sepsis?" *Center for Disease Control and Prevention.* (2019)

    https://www.cdc.gov/sepsis/what-is-sepsis.html

5. Dunigan, Christina. "Legal Abortion Death: Marla Cardamone, 18" *Clinic Quotes.* (2012)

    https://clinicquotes.com/legal-abortion-death-marla-cardamone-18/

THE ABORTION PILL RU486

Chapter 13

Mifepristone Abortions & Trauma Women Experience

I grabbed a towel to bite on in order to keep from screaming and was nearly passing out. As I got up I saw blood everywhere. I saw parts of my baby, images I will never be able to erase. I fell to my knees in pain and was blacking out. Concerned that the guys would see all the blood and clumps, I got on my knees and cleaned it up.

Ann[1]

In this quote, Ann is writing about her experience with the abortion pill, also known as RU486, mifepristone, or medical abortion. RU486 is the brand name for mifepristone. Ann's story is not unique. I've read countless stories from women just like her. Unlike surgical abortions, the abortion pill causes the abortion while the woman is alone at home or at a hotel room. Most surgical abortions are relatively quick. You simply go to the clinic and let the abortionist do the hard part. But with the abortion pill, there is no abortionist there to do the hard part. You have to clean it up yourself. Abortion clinics instruct women to have the abortion on a toilet bowl and flush. This has caused some people to start referring to these abortions as toilet boil abortions.

In order for the abortion pill to be successful, everything must be passed out of the uterus. We are going to look more at Ann's story and the downsides to the abortion pill that abortion clinics rarely warn about. But first let's look at how the pill causes an abortion.

An abortion with pills actually requires two different types of pills. The first pill is mifepristone, which most people know by the brand name RU486. This is the pill that blocks progesterone and causes the death of the embryo. The second pill is Misoprostol. This pill is taken 24-48 hours after to cause contractions to expel everything out of the uterus.

To understand how RU486 kills the embryo, it is important to remember some of the biology from chapter 5 where we learned about the role the corpus luteum plays in producing progesterone. It is also important to remember chapter 8 where we learned that soon after implantation the blood of the embryo and the blood of the mother are able to exchange nutrients in order to grow and support the embryo.

This is done when the embryo's blood in early placenta tissue called the trophoblast comes in close contact with the mother's blood in the lining of the uterus. We also know that the follicle that ovulated the egg turns into a gland called the corpus luteum which produces progesterone. The word progesterone simply means "pro-gestation" and is key to gestating the embryo.[2] The progesterone thickens the lining of the uterus and keeps that lining nice and healthy to support the embryo. In order for the lining to benefit from the progesterone, there are receptors in the uterus that must receive that progesterone. The embryo in return produces a hormone called hCG which causes the corpus luteum to continue producing progesterone. Without an embryo and hCG, the progesterone stops. The lining deteriorates or breaks down and is expelled as menstrual blood.

RU486 interrupts the process. This drug essentially tricks the progesterone receptors in the uterus. The receptors receive the RU486 as if it were progesterone. But the RU486 doesn't activate the receptors to support the lining of the uterus. The progesterone can't be received because the RU486 has already taken up those receptors. This is why RU486 is called a progesterone blocker. It literally blocks the progesterone from the receptors and renders the receptors useless. Without the progesterone, the lining begins to deteriorate and break down the same as it would if the mother were not pregnant. Only this time she is pregnant.

What this means is that the embryo's blood is no longer able to exchange nutrients and gasses with the mother's blood. As the cells of the lining die off, the placenta tissue of the embryo is separated from the lining of the uterus.[3] The blood of the embryo and the blood of the mother are no longer in close contact. The embryo can't expel waste and it can't receive nutrients and oxygen. This is what causes the death of the embryo. In essence, the embryo dies of neglect. Neglect in a parent-child relationship is when the parent fails to provide for the basic needs of the child. Failing to feed a child a minimally nutritious diet, failing to provide a safe living environment, and failing to clothe a child are examples of neglect. In an RU486 abortion, the mother fails to provide for her embryo the necessary nutrition and also fails to provide the necessary environment in the uterus.

It is important to understand that death through neglect is still an act of violence. Some people see RU486 abortions as less morally objectionable. Many pro-life people seem less motivated to end RU486 abortions than surgical abortions. But killing is always a violent act regardless of how the killing occurred. I think some of the confusion in this area is due to a misunderstanding of types of violence. Killing is an act of violence but so is significant bodily harm. Some violent acts are killing. Some violent acts are not killing but are acts of bodily harm, such as breaking an arm or even giving someone a black eye. Many violent acts are both. This is true of surgical abortions. The further developed the fetus becomes, the greater the acts of bodily harm that are done in the process of killing that fetus. And so late-term second and third trimester abortions are rightfully seen as the most violent. For example, the crushing of the skull is not typically done in the first trimester. But it is done later in pregnancy. The act of crushing the skull is increasingly violent. RU486 abortions are not doing direct physical trauma to the body of the embryo in order to kill it. But these abortions are still causing the intentional death of the embryo, which necessarily means that all abortions are acts of violence, even the very early ones. It is important to understand that every abortion necessarily kills a prenatal human. If abortion didn't kill, it wouldn't be an abortion. If abortion didn't kill, there would be no pro-life movement. This is why abortion is fundamentally different from any other medical practice.

Once the embryo has died, its body, along with the uterine lining, deteriorates, making it easier to expel from the uterus. This is the reason RU486 is used. It is not necessary to block progesterone in order to expel the embryo from the uterus. Some abortionists have been known to skip the RU486 entirely and simply induce labor to deliver the embryo. But the abortion industry has found that it is much easier to expel the embryo after its body, along with the uterine lining, has deteriorated. It requires weaker contractions to expel and there are fewer complications.

The second pill used in RU486 abortions is Misoprostol. This is the pill used to cause cramping, contractions, and labor. When a woman has her period, she naturally has very small contractions or cramping. RU486 causes the uterus to attempt a normal period including the cramping. It is not a normal period, however, because the woman must pass the dead embryo. And it is not like a normal period because the Misoprostol induces contractions that are stronger than a normal period. Misoprostol is also used in labor and delivery wards as another way to induce labor. It is known in labor and delivery for causing very strong contractions. Some even describe the contractions as violent. The important thing to remember is that it induces labor. It is not just like a heavy period. And it is not like having an early miscarriage, unless your doctor gives you something like Misoprostol to help you pass the miscarried child. This is important to understand because abortion clinics frequently downplay and mislead women about the amount of pain and blood they will experience. If a woman knows how much pain and difficulty she is signing up for, she may choose a surgical abortion instead or she may even choose not to have an abortion. Planned Parenthood describes it on their website as "like having a really heavy, crampy period" and "very similar to an early miscarriage."[4]

Ann's story is typical of the countless stories I've read about women having RU486 abortions with Misoprostol. She describes the Misoprostol as follows:

> *It was time for me to take the last dose. As I put the pills in my mouth and let them dissolve, within ten minutes I started to feel intense cramps. When the cramps became unbearable I made my way to the bathroom. I locked the door and experienced the most severe pain I had ever felt in my life. I sat on the toilet and bent over in pain. I wanted to scream but my ex-boyfriend and his friends were right outside the door in the living room watching TV. It was a small apartment. I grabbed a towel to bite on in order to keep from screaming and was nearly passing out. As I got up, I saw blood everywhere. I saw parts of my baby, images I will never be able to erase. I fell to my knees in pain and was blacking out. Concerned that the guys would see all the blood and clumps, I got on my knees and cleaned it up. As soon as I left*

the bathroom I was about to faint when my ex-boyfriend helped me to bed."[5]

The pain and blood can be very traumatic and unexpected but are not the only traumatic factors. Seeing the body parts of the embryo after it has passed can also be traumatic. Abortion clinics don't typically warn women about what they will see although a few clinics may urge women to flush without looking. RU486 abortions can be done up to ten weeks gestation. This is the point when we begin calling it a fetus due to the fact that it clearly looks like a baby. When having this abortion, women can see clearly the features of the child, including fingers, toes, face, umbilical cord, and placenta. Sometimes the child comes out intact. Sometimes she comes out in pieces. The child may or may not be inside the gestational sac. You will remember that embryonic folding occurs by the sixth week. The child is about the size of a pea at six weeks and as large as a strawberry by ten weeks. As you can imagine, passing a strawberry-sized body through the cervix would take significant contractions. Picking that child up and placing it in the toilet to flush it must create a heavy emotional load.

It's important to mention the men as well. Abortion certainly has an emotional impact on the fathers. These men are often the most forgotten in the abortion issue. Some men want their babies and suffer greatly when they find out that their children were aborted and they were powerless to stop it. Other men carry the guilt of coercing her, manipulating her, or paying for her abortion. Abortion has a traumatic effect on all involved: the abortionist, the mother, the father, and even the rest of the family.

In 2016 the FDA approved significant changes to the protocol for RU486 abortions, which had the effect of expanding RU486 abortions and making them more traumatic for women. The change did two things. First, it increased when you can get this abortion from seven weeks gestation to ten weeks gestation. Second, it decreased the dosage of RU486 and increased the dosage of misoprostol.[6] By increasing the gestation to ten weeks, the FDA dramatically increased the number of abortions that could be done with the pill. It also means that women are passing significantly larger fetal bodies. At ten weeks, we are talking about a strawberry-sized body. The body parts are much larger and more distinct. The woman likely will see the body, possibly intact or possibly in pieces.

The change in dosage also increases the trauma for women. The RU486 dosage was decreased from 600 milligrams to 200 milligrams, a third of the dose. This was a big win for abortion clinics because RU486 is an expensive drug. A third of the dose equals a third of the cost for the abortion clinic to buy that drug. It also means that the lower dose is less successful in killing the embryo. If a woman only takes the RU486 and doesn't finish with the misoprostol, there is a significant chance the embryo could survive and continue with no birth defects. The dosage of misoprostol increased from 400 milligrams to 800 milligrams with the possibility of taking additional misoprostol. And so now the chances of the embryo being loosened from the lining are less and the contractions needed to expel the embryo are much greater. Doubling the dosage of misoprostol means stronger contractions and more pain. All of these factors are combined to make RU486 a more traumatic experience for women.

Some of the more common side effects of RU486 are nausea, vomiting, diarrhea, and headaches. Sometimes heavy bleeding can last for weeks.[7] The more severe complications include what is called an incomplete abortion. Incomplete means that parts of the child or supporting parts are still in the uterus. These parts will result in an infection which can become very dangerous and even life-threatening. If a woman doesn't pass everything, then she must go back for a surgical abortion to finish the job. The story of Ann which we looked at earlier in this chapter was an incomplete abortion. She commented, "As they performed the D&C I couldn't help but think my baby was a fighter."[8] Another serious complication is an ectopic pregnancy. RU486 abortions only work if the embryo is in the uterus. An RU486 abortion on an ectopic pregnancy is dangerous. Some other more serious complications include interactions with other drugs, allergic reaction to the drugs, and the presence of an IUD. RU486 can in some rare cases cause death. According to the FDA, 24 women have died from RU486 complications in the United States between September of 2000 and the end of 2018.[9] While abortion supporters emphasize that these deaths are rare, the deaths are real. About one or two women typically die in the United States each year from RU486.

The FDA has regulations in place to try to reduce the risk of complications and death. These regulations are called REMS, which stands for Risk Evaluation and Mitigation Strategy. The FDA requires that doctors be certified in the REMS program in order to dispense the drugs. In other words, not just any doctor can write a prescription for RU486. The doctor has to dispense the RU486 in a clinic or hospital setting. You can't get this drug at your local pharmacy. In order to get certified, the doctor has to meet minimum competencies, such as the ability to do surgical abortions and diagnose ectopic pregnancies. The doctor must also have access to certain equipment, such as the equipment necessary to provide blood transfusions.[10]

Not surprisingly, the abortion advocates are fighting hard to do away with REMS and any other restrictions designed to protect women. The ultimate goal is to make RU486 as easy to get as possible, including over the counter. As an over the counter drug, there would be no supervision by a doctor. None of the precautions in REMS would be in place. The abortion industry is trying to expand RU486 in some other ways on the road to over the counter abortions. One that many people have heard of is the so-called webcam abortions which are done via telemedicine. Planned Parenthood of Iowa first rolled out a model where the patient would go to an abortion clinic, but the doctor would not be on site. Instead the doctor would interact with the patient and clinic staff over webcam. Many states have now banned this practice but a few clinics still do these webcam abortions. The latest FDA approved trial that is now being conducted is an at-home webcam abortion. In this trial, the interaction with the doctor would be in your own home and the pills would be mailed to you.

Webcam abortions are ultimately a step toward the eventual goal of over the counter abortions. Abortion advocates call this "self-managed abortion." Ironically, Roe v. Wade was supposed to get us away from dangerous "self-managed abortions." After all, isn't this why pro-choice activists are running around waving coat hangers? But with RU486, self-managed abortions are making a comeback. RU486 has drastically changed the abortion landscape. Regardless of the fact that the FDA prohibits self-administered RU486 abortions, some women still do them. It is surprisingly easy to buy black market pills. You don't even have to go to the dark web or a sleazy drug dealer to buy them, although some people may do that. People buy black market pills online. It's not safe and the FDA warns people not to do it. But people do it anyway. This is likely the future of illegal abortions. It is already happening. When Roe v. Wade is overturned and some states prohibit most abortions, RU486 will likely be the illegal abortion of choice. But this doesn't stop pro-choice activists from waving around coat hangers and making themselves look silly.

The abortion pill has drastically changed the way we do abortions in the United States and around the world. It has allowed women to see the humanity of their aborted children in a way that they couldn't with surgical abortion. It has made the abortion procedure longer, more painful, and more traumatic. But one last change that we must look at is the way it has made abortion more profitable. RU486 has been a gift to the abortion industry. Not only can they typically charge more for RU486 than for a surgical abortion, they can save on a lot of the costs as well. There is less equipment and overhead involved. The abortion itself only requires the woman to take a pill before she walks out of the clinic. The clinic doesn't have to prepare the woman for surgery. The doctor doesn't have to take the time to do the surgery. It's super easy. In fact, the overhead for RU486 is so much lower that there are abortion clinics all over the country that only do RU486. It's much easier and less expensive to open a new abortion clinic if it only offers RU486.

Another benefit to the abortion industry is that the abortionist doesn't have to deal with the complications should they arise. The complications from surgical abortions can be very costly for the abortionist and the clinic. An abortionist may have to deal with complications like uterine perforation and hemorrhaging himself in the clinic. If he has to send the woman to an emergency room, he would have to face the negative publicity of an ambulance at his clinic. He may also face the risk of medical malpractice lawsuits from trying to treat complications in the clinic. With RU486, the clinic can direct the woman to go to the emergency room if necessary and let her complications be some other doctor's problem. There is no ambulance at the clinic. There is no 911 call from the clinic. There is no bloody mess to clean up in the clinic. It is a really great benefit for the abortion clinic.

The great losers in RU486 abortions are the unborn children and their mothers. I can't imagine the horrific experience of being alone, experiencing contractions, passing my perfectly formed little child, and flushing that toilet. These women deserve to at least know the truth before they decide to do this abortion. They deserve all the details, including how far the child is developed, the severity of the contractions, and what they might see and experience. At a minimum, they deserve the truth.

Chapter Notes

1. "Ann" Taken from Facebook on March 27, 2019. Ann's real name has been withheld to protect her privacy. Lightly edited for grammar and readability.

2. Paul, Maureen, et al. *Management of Unintended and Abnormal Pregnancy* Hoboken: Wiley-Blackwell, 2009, pp. 113

3. Ibid, pp. 113

4. "The Abortion Pill" *Planned Parenthood.*

   https://www.plannedparenthood.org/learn/abortion/the-abortion-pill

5. "Ann"

6. Rafie, Sally. "Abortion Pill Label Change: What Pharmacists Need to Know" *Pharmacy Times.* (2016)

   https://www.pharmacytimes.com/contributor/sally-rafie-pharmd/2016/04/abortion-pill-label-change-what-pharmacists-need-to-know

7. "Medical Abortion" *Mayo Clinic.* (2018)

   https://www.mayoclinic.org/tests-procedures/medical-abortion/about/pac-20394687

8. "Ann"

9. "Mifepristone U.S. Post-Marketing Adverse Events Summary through 12/31/2018" U.S. *Food and Drug Administration.*

   https://www.fda.gov/media/112118/download

10. "Prescriber Agreement Form: Mifepristone Tablets, 200 mg" *U.S. Food and Drug Aministration*

    https://www.accessdata.fda.gov/drugsatfda_docs/rems/Mifepristone_2019_04_11_Prescriber_Agreement_Form_for_GenBioPro_Inc.pdf

ABORTION PILL REVERSAL

Chapter 14

Progesterone Support to Reduce the Effects of RU486

"I wanted so badly to have this baby and to have a second chance."

Rebekah Buell-Hagan[1]

After RU486 was approved by the FDA for abortion in 2000, something interesting happened. Woman started coming forward after taking the first pill and changing their minds about going through with the abortion. Sometimes people make impulsive decisions and end up regretting those decisions. This is especially true when a woman can walk into an abortion clinic and in almost no time at all, ingest a pill to cause an abortion. She may impulsively go to an abortion clinic when she first discovers that she is pregnant and emotions are running high. But after she takes the first pill at the clinic, she may put some more thought into it and realize that this isn't the decision she wants to make. This is exactly what happened in 2007 when a lady in North Carolina approached her family doctor after changing her mind. Her doctor, Dr. Matthew Harrison, did some research on how RU486 blocks progesterone and decided to try an idea. He injected her with progesterone that he happened to have in his office. He knew it was a long shot and warned her that it may not help. Six months later she became the first recorded instance of a woman delivering a healthy baby after taking RU486 and then progesterone in an attempt to save the child.[2]

On the other side of the country, another doctor named Dr. George Delgado had a similar experience giving a patient progesterone and saving her child. As word spread of this case, Dr. Delgado began receiving more and more calls from people wanting to save their unborn children after taking RU486. Dr. Delgado, Dr. Harrison, and another provider named Dr. Mary Davenport worked together to start a nationwide program to administer progesterone to women who regretted their decision to take RU486.[3] Today that program is called Abortion Pill Rescue and has a nationwide network of hundreds of providers that prescribe and administer progesterone. A nationwide network of pro-life pregnancy care centers administers a 24/7 hotline to connect women with providers. You can learn more about the network at www. AbortionPillReversal.com.

The idea of administering progesterone for pregnancy is not a new one. For decades fertility doctors have been giving progesterone regimens to patients to help with fertility and reduce the risk of miscarriage. Anyone who has gone to a fertility specialist is likely to be familiar with progesterone support. It is one of the more memorable treatments as it usually includes a painful injection into the muscle. Progesterone is typically administered three different ways: an intramuscular injection, orally, and with a suppository. The progesterone suppository is like a large pill that is inserted into the vagina. A doctor may prescribe a combination of the three to get the desired dosage. While the term "progesterone support" is used in fertility clinics, doctors trying to reverse an RU486 abortion use the term "abortion reversal." But these terms both refer to administering progesterone in order to prevent the loss of the embryo or fetus. I prefer the term "progesterone support" as it is more descriptive of the specific medical care that is being provided.

It shouldn't be surprising that progesterone support to undo RU486 is controversial. The abortion industry is quite vocal with their opposition to progesterone support. If a woman changes her mind and calls the abortion clinic, she will typically be told that she has to take the second pill and finish the abortion. Women are strongly discouraged from attempting progesterone support. Some abortion clinics will even go so far as to falsely claim that RU486 will give the baby birth defects. [4,5] This is despite the fact that there is zero evidence that RU486 causes any kind of defects or deformities. Even the American College of Obstetrics and Gynecology, which is vocally pro-choice and vocally opposed to abortion pill reversal, in a 2017 paper, said that RU486 "is not known to cause birth defects."[6]

But the most widely made and unfounded claim by critics is that progesterone has too many possible side effects and is dangerous to women. But progesterone support is widely accepted as extremely safe and is a common treatment both for infertility and for women going through menopause. It is true that any medication can carry risk. But the risks associated with progesterone support aren't substantial. And yet this is a common claim that is unfortunately used to scare women.

Those critics of progesterone support who are more objective and honest about the medicine will tell a woman who has changed her mind that there is some chance her baby could survive. Even without progesterone support, there are two possible steps a woman can take to give her embryo/fetus a chance at survival. If it has been less than an hour after she took the first pill, she could induce vomiting or have her doctor induce vomiting to attempt removal of the RU486 before all of it has been digested. A woman would have to act fast for this to be possible. The other step is to simply refuse to take the second pill and hope for the best. There is a lot of disagreement as to the chances of an embryo surviving only the first pill. But the chance of survival could be as high as 50%.[7] It is also noteworthy that in 2016 the FDA approved a significantly lower dose of RU486. It is now recommended that the first pill have 200 milligrams instead of 600 milligrams. The lower dose could mean a greater chance of survival if the woman doesn't take the second pill.[8] Most women, however, are not informed that these two options exist. If she were to call her abortion provider and express a desire to continue the pregnancy, it is not likely that she would be informed of the possibility of continuing. It is more likely she would be told that she must take the second pill and finish the abortion.

Most of the controversy surrounding progesterone support to reverse RU486 is over whether or not the progesterone increases the likelihood of a woman being able to save her baby. The critics, who reside mostly in the abortion industry, argue that added progesterone does not increase the patient's chances of continuing the pregnancy. They further claim that doctors doing it are giving women false hope. The doctors doing the progesterone support claim that it does increase the patient's chances of continuing the pregnancy and that the abortion industry isn't giving the patient enough hope that it can be done.

In 2018, Dr. Delgado and his network published a case study showing the results of their work. It analyzed the results of 547 women over four years and using different types of progesterone support. He calculated the survival rate and published the results. He found an overall survival rate of 48%, meaning that 48% of the women using progesterone support were able to carry their babies to live birth. He contends that the survival rate would have only been 25% without the progesterone according to previous research. And so he is claiming that the progesterone significantly increases the chances of live birth. The critics are now claiming that they don't accept a 25% survival rate without progesterone. They now think its 50% and therefore the progesterone didn't help. In my opinion, the critics are just moving the goal posts. What's interesting about the study is that the thirty eight women receiving the highest amount of progesterone, six or more intramuscular injections, were the women most likely to have a live birth. They had a survival rate of over 89%.[9] And so how does the chance of survival increasing with higher dosages if the progesterone support doesn't work? If the progesterone support doesn't work, we would expect about the same survival rate regardless of the dosage. But if the progesterone support does work, then it makes sense that greater support would mean higher survival rates. In my opinion, the higher survival rate is compelling evidence.

It is important to understand the difference between a case study and a randomized trial. A case study is simply reporting on the results of the treatment. In other words, it only reports what the doctor did for his patients and the results. Published case studies are evidence and are worthwhile. But randomized trials are far more compelling. A randomized trial is much more rigorous and compares those who received the treatment to those who received a placebo or didn't receive treatment. Dr. Delgado has acknowledged that a randomized trial is needed. But he won't do the trial because it would be unethical to give a placebo to a woman who wants to save her baby.[10]

As of the writing of this book, an abortionist named Dr. Mitchell Creinin is attempting a small randomized placebo-controlled trial that will include forty women. Presumably the women he will recruit are patients at his abortion clinic. His ethical justification is that all of these women are seeking abortions anyway. He will give some of these women progesterone and others a placebo in order to analyze how many of their embryos are still alive two weeks after they took the RU486. After the two weeks are up, he will do surgical abortions on any remaining living embryos and fetuses.[11] And so he will be experimenting on these embryos and possibly saving some of these embryo/fetuses with progesterone, only to surgically abort them after they no longer benefit his research.

It's important to understand the background of Dr. Creinin and the organization funding his study. Creinin is an abortionist's abortionist. He is widely recognized as an expert. He was an expert witness when congress passed the Partial Birth Abortion Ban and is a co-author of *Management of Unintended and Abnormal Pregnancy*, the primary textbook used to train abortionists. He is also known to have been cited by the FDA for violations in a previous study.[12] Most notably, Creinin is already a vocal critic of abortion pill reversal which begs the question of whether or not he can do the research objectively and not allow his biases to skew the research.[13] A further problem for this study is the organization paying for it. Society of Family Planning is a pro-choice organization seeking to expand abortion, especially RU486 abortions. This organization spends millions of dollars each year in research grants with the bulk of the money going toward expanding RU486 as widely as possible. In 2018, over eight million dollars in grants were awarded to expand RU486 in areas such as mail-order abortions, self-administered abortions, over the counter abortions, and abortions in family practice.[14] All this is to say that this small randomized trial is being conducted with an agenda. Not only must Dr. Crenin do the research, but he must also be able to convince the public that his research is legitimate.

Editors Note: The Crenin study was cancelled and annouced after the publication of this book.

Chapter Notes

1. "One Young Woman's Abortion Pill Reversal Story" EWTN. (2018)

   https://www.youtube.com/watch?v=TWyf6Y_BPlQ

   Quote taken from interview at 2 minutes 30 seconds Ms. Buell-Hagan was a successful abortion pill reversal patient and is now an advocate for reversal.

2. Cleveland, Margot. "Are Abortion Reversals Science or Scam?" *The Federalist*. (2017)

   http://thefederalist.com/2017/05/16/abortion-reversals-science-scam/

3. Ibid

4. Devine, Daniel. "Cynthia's choice" *World Magazine*. (2013)

   https://world.wng.org/2013/04/cynthias_choice

5. Hobbs, Jay. "On Abortion Pill Reversal, It's Time to Hear from the Women" *Pregnancy Help News*. (2018)

   https://pregnancyhelpnews.com/on-abortion-pill-reversal-it-s-time-to-hear-from-the-women

6. "Facts Are Important: Medication Abortion "Reversal" Is Not Supported By Science" *The American Congress of Obstetricians and Gynecologists*. (2017)

   https://www.acog.org/-/media/Departments/Government-Relations-and-Outreach/FactsAreImportant-Medication AbortionReversal.pdf

7. Gordon, Mara. "Controversial 'Abortion Reversal' Regimen Is Put To The Test" *National Public Radio*. (2019)

   https://www.npr.org/sections/health-shots/2019/03/22/688783130/controversial-abortion-reversal-regimen-is-put-to-the-test?fbclid=IwAR1n613V0tWOTZyENqXYtHNqyiY0aZtBEc_xa0f3qzNIdnRL_InA98DWK9A

8. Rettner, Rachael. "Abortion Pill Gets New Label: 5 Things to Know About Mifepristone" *Live Science*. (2016)

https://www.livescience.com/54238-abortion-pill-mifepristone-label.html

9. Delgado, George. et al. "A Case Series Detailing the Successful Reversal of the Effects of Mifepristone Using Progesterone" *Issues in Law & Medicine*. Volume 33, Number 1 (2018)

   https://issuesinlawandmedicine.com/wp-content/uploads/2018/10/Delgado-Revised-09-2018-1.pdf

10. Ibid

11. "SFP research grant awards" *Society of Family Planning*. (2018)

    https://www.societyfp.org/Research-and-grants/Grants-funded.aspx

12. Spears, Larry. "Warning Letter" *Food and Drug Administration*. (2002)

    http://abortiondocs.org/wp-content/uploads/2014/03/Creinin-Mitchell-FDA-Warning-Letter-6-12-2002.pdf

13. Gordon

14. "SFP research grant awards"

THE FORCEPS

Chapter 15

Forceps Abortions & Aborting Disabled Fetuses

Then I inserted my forceps into the uterus and applied them to the head of the fetus, which was still alive...

Late-Term Abortion Specialist Dr. Warren Hern[1]

When I first saw the type of forceps used for abortion, it was an instrument that overwhelmed and disturbed me more than any other instrument. Most people find the abortion forceps to be the most upsetting. This instrument looks much more overtly violent than the curettes that I had seen previously. Seeing this instrument is a turning point for many people. It's the moment when you realize that someone obviously designed this instrument with death and destruction as its goal.

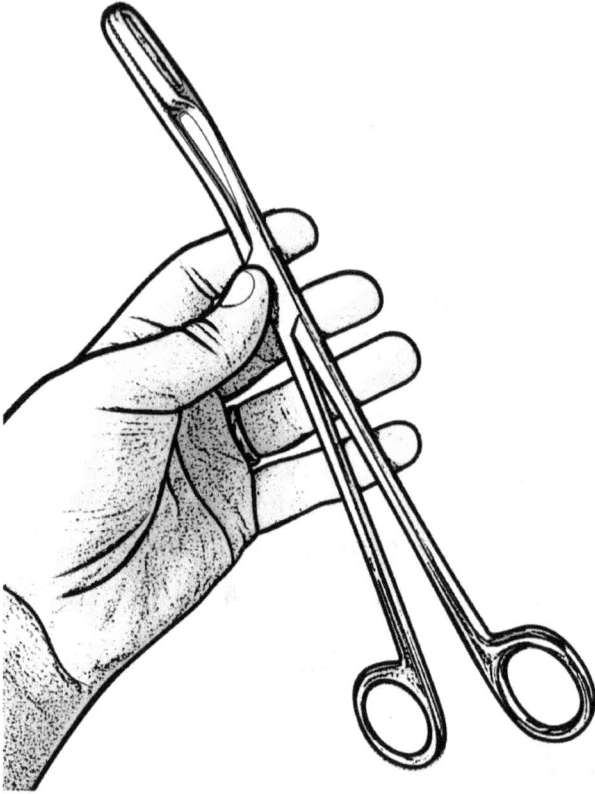

Sopher Ovum Forceps

The forceps used in forceps abortions are not like those used to aid the delivery of babies. Forceps come in many different shapes and sizes depending on the task for which they were designed. The forceps used for delivering babies have large hoops on the end that wrap around the baby's head. Some have described it as looking like large salad tongs. It doesn't look nearly as menacing. These are not the forceps used in abortion. The only time this type of forceps would be used in an abortion would be for an abortion near the end of pregnancy where the dead baby is delivered whole and the forceps are needed to aid delivery.

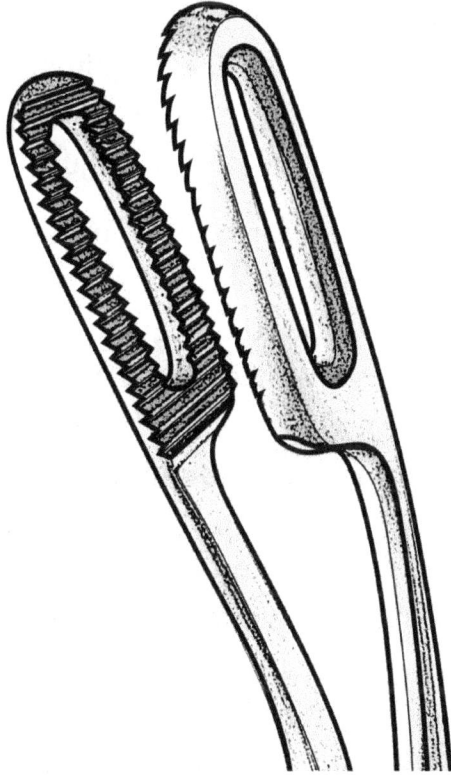

The Jaws and Teeth

The forceps used for abortions instead have smaller hoops or jaws that are lined with sharp teeth. It looks a lot like a needle nose pliers except that the teeth are larger and sharper. The jaws are about the size of a large thumb. Abortion forceps come in many shapes and sizes as well, but they all have several characteristics in common. They are all made of rigid metal. They all have handles and a hinge in the center allowing the doctor to open and close the jaws. And they all have the smaller hoops lined with sharp teeth. Forceps may have other features as well, such as a ratchet. The ratchet is used to keep hoops closed tight until the doctor is ready to open them. This way an abortionist can grab a body part and it won't let go.

While some forceps used in abortion aren't necessarily designed for abortions, many of them are. One such forceps is the Hern Ovum Evacuation Forceps.[2] This forceps takes its name from Dr. Warren Hern, one of the most well-known and outspoken late-term abortionists in the country. Hern's practice in Colorado performs abortions well into the third trimester. The word "ovum" refers to the egg. And the word evacuation simply refers to removing something from the uterus. I don't know who had the idea to call it an "ovum evacuation" forceps instead of a fetus evacuation forceps. The forceps is never used to remove an egg, fertilized or unfertilized, from the uterus. It isn't even used to remove embryos. It is only used to remove fetuses at the very end of the first trimester and later. The Hern forceps is made specially with the shape and size he prefers. I can't imagine being such a prolific killer as to have a deadly instrument made exactly to my specifications.

The forceps are generally used in one of two ways: to remove pieces or to crush. To remove pieces is relatively straight forward. You simply grab whatever body part you can find, an arm or leg for example, and pull it off. You must continue removing pieces until all that is left is the head. This method of pulling off body parts is typically done in the second trimester. In the first trimester, the body is small enough as to not need a forceps. And in the third trimester, the skeletal system has hardened to the point where it is very difficult if not impossible. However, if the fetus is killed with a lethal injection the day before, decomposition may aid in removing pieces.

Most of these abortions occur in the second trimester when it is too large to be vacuumed out but not too large to remove in pieces. In most second trimester forceps abortions, the fetus dies when it bleeds out as a result of losing limbs. It dies in much the same way that you or I would die if our limbs were removed. In the second trimester, it is hard work to remove pieces. It takes substantial arm strength to get the job done. The sharp teeth help to get a good grip. When all that is left is the head, then the head must be crushed. The head is the greatest challenge as it is large and round. One way to deal with this problem is to crush it with the forceps. Then the head that was crushed must be scraped out with the sharp uterine curette. The crushing results in sharp skull fragments. These skull fragments must be removed carefully so as not to damage the uterus. Removing sharp skull fragments is risky.

In addition to removing pieces, the forceps are also used for crushing. Not only are they used for crushing the head but also the rest of the body. The sharp teeth of the forceps are especially important when crushing. With this method, the abortionist simply crushes up the body parts till they are nearly unrecognizable and then pulls them out. One of the downsides of crushing is that it doesn't preserve body parts which can then be legally donated or illegally sold for research. The perceived benefit of crushing is that the abortionist and staff don't have to look at the recognizable body parts. It allows the abortionist to dehumanize the child so that he doesn't have to confront the violence so directly. It also requires less strength to remove pieces if they have been crushed up first.

Another downside to crushing with the forceps is that it is harder to confirm that the abortionist removed everything. Normally the clinic staff has to examine the parts with a backlight to make sure they got everything.

If parts are left in the uterus, it can become infected and further damage can be done to the uterus. This is especially true if pieces of the skull are left behind. Skull fragments are sharp. Sharp objects in the uterus can become dangerous very quickly. One technique used to overcome the problem of examining tissue after it has been crushed is weighing the tissue. The abortionist uses measurements from an ultrasound to estimate the total weight. If the parts don't weigh enough, then he suspects that there may still be pieces inside.

Remember, forceps abortions are mostly done in the second trimester. Sometimes they are done into the third trimester, but by then the baby has grown very large and the bones have hardened further. Using the forceps in the third trimester is often too difficult and requires a lethal injection abortion which is described in the next chapter. As you now know, the fetus in the second trimester is a fully developed baby. All the parts are in place, including organ systems, face, hands, and feet. The only difference between a second trimester fetus and a third trimester viable fetus or even a birthed newborn baby is size and maturation. All that needs to happen for a second trimester fetus to be born is simply to get bigger and become more mature.

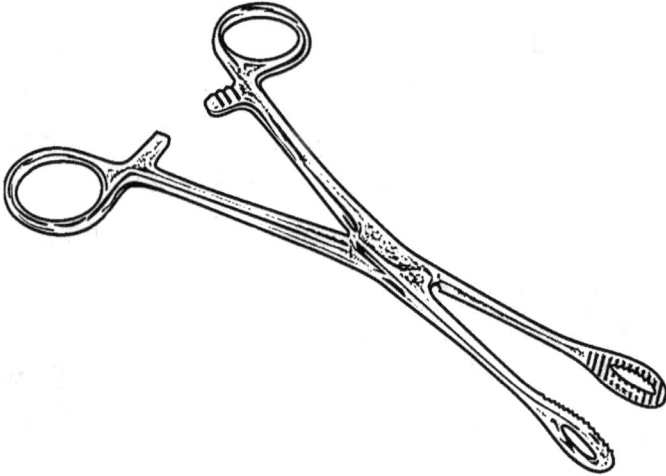

Sponge Forceps

The earliest that a forceps may be used is at the end of the first trimester. When doing a vacuum aspiration abortion, the doctor might find that the head is too big to be vacuumed. He may use a sponge forceps to crush the head and remove it. The sponge forceps is the smaller cousin to the larger forceps used in second trimester abortions. The forceps can be used from the end of the first trimester right up to full term.

Women get abortions for nearly every conceivable reason. Planned Parenthood v. Casey, the Supreme Court decision that replaced Roe v. Wade in 1992, declared abortion a constitutional right for any conceivable reason through 24 weeks gestation. This has been one of the conflicting issues in public opinion.

While half of the country claims to support Roe and its companion cases, most Americans do not agree that you should be able to get an abortion for any reason. Most Americans do think that there should be a justifiable reason. But Roe has made it impossible to even have a public discussion over what justifies an abortion.

This has resulted in two terms that we frequently hear: elective abortions and therapeutic abortions. These terms are widely used in public debate but are also necessitated by medical billing. If your insurance only pays for abortions for certain reasons, you need to have billing codes that allow the doctor to bill correctly for the given situation. For example, Medicare usually pays for abortions only in the cases of rape, incest, or a qualifying medical reason. Elective abortions are generally understood to be those that are done for reasons other than medical reasons. If a woman gets an abortion because she doesn't feel like having a baby, that is an elective abortion. She had a reason for making that decision, but it was not a medical reason. But if a doctor recommends an abortion for a medical reason, that is considered to be a therapeutic abortion. The line between the two, however, is fuzzy at best. What if a doctor recommends abortion because the woman is stressed out about having a baby? Stress could be considered a mental health condition and have negative health consequences. Something as minor as stress could be used as an excuse to label it a therapeutic abortion. In fact, some abortion advocates go so far as to claim that every abortion is potentially saving the woman's life and therefore every abortion is therapeutic. My position and that of many people in the pro-life movement is that abortion is a horrifically violent act and only justifiable when the mother is given necessary lifesaving treatment that may inadvertently kill the unborn child. The most common example of this is surgery for ectopic pregnancy, as explained in chapter 5. These lifesaving treatments are not done in abortion clinics.

The reasons given for abortion change somewhat once we get into the second trimester and the forceps are used as the primary instrument. This is because the number of fetuses aborted for "fetal abnormalities" increases somewhat later in the second trimester. The term "fetal abnormality" is a euphemism for any fetus with a disability or medical condition. These children are labeled abnormal and often aborted as a result. It is an Orwellian euphemism, as are so many used by the pro-choice movement. We would never say that an adult with Down Syndrome has "adult abnormalities" or that a teen with autism has "teen abnormalities." But any unborn child with a medical condition is labeled abnormal and abortion is offered as the solution. These wide ranging medical conditions and disabilities are often discovered with an ultrasound around 20 weeks gestation. This is one of the reasons why the pro-choice movement fights so hard against 20 week abortion prohibitions. They don't want to lose the ability to abort an "abnormal" baby.

These medical conditions can range from very serious conditions that will inevitably kill the baby to very minor conditions that can be managed or fixed with surgery. Some of these are considered "chromosomal abnormalities" because the chromosomes did not fuse together correctly at conception. Down Syndrome or trisomy 21 is a common example. The 21st chromosome pair has a third chromosome attached causing Down Syndrome. This occurs at conception. Other disabilities are in a category called neural tube defects. These birth defects, which include spina bifida, cleft palate, and club feet occur at six weeks gestation when the embryo does not fold up completely. These are just two categories of disabilities for which people choose to abort. For these babies, instead of being given the dignity of medical care and a natural death, they are violently euthanized with forceps.

The pro-choice movement often jumps to the hard cases, those unborn babies with a terminal condition, as if a violent forceps abortion is better than a natural death. But many of these babies aborted because of disabilities are not terminal. Whether it's spina bifida or Down syndrome, with adequate medical care and therapy, these children can go on to live very happy and fulfilling lives, so long as they aren't aborted first. Two thirds of Down syndrome babies are aborted in the United States. Of those diagnosed with Down syndrome in utero, 92% are aborted. And yet over 95% of teens and adults with Down syndrome reported that they liked how they look, liked who they are, and were happy with their lives.[3] Almost all of these aborted Down syndrome babies would have grown up to be happy, fulfilled, and differently-abled adults.

Unfortunately, the tests that are used to diagnose babies with Down syndrome in utero put the child at risk of miscarriage or still birth. Parents are not generally encouraged to get the test unless they are willing to abort. That is why 92% of those diagnosed with Down syndrome are aborted. Those who would not abort are not likely to put their baby at risk of miscarriage or still birth with the test. The tragic result is that many healthy babies that do not have Down syndrome die from the test each year. The most common diagnostic test is called amniocentesis with approximately 200,000 tests done each year in the

United States.[4] The parents typically first do a blood test and ultrasound to determine if their baby is at a higher risk of Down syndrome. Amniocentesis is usually only done when there is a higher risk and when the parents are considering abortion. With the risk of miscarriage at 1 in 200 to 1 in 400, that means that approximately 500-1,000 babies are miscarried due to amniocentesis each year in the United States.[5] One published case study found that 93% of the babies that underwent amniocentesis at one clinic were found to have normal chromosomes.[6] If these numbers hold true, as many as approximately 900 chromosomally normal babies are lost to miscarriage and still birth in the United States each year so that their parents can keep open the option of abortion. This shows the wide ranging affects that abortion culture has had on our society.

Even with the increase of abortion of disabled babies in the second trimester, most of these abortions are still purely elective according to late-term abortionist Martin Haskell.[8] In this chapter we've looked at the increasingly violent abortions done generally in the second trimester with the forceps. In the next chapter we will look at lethal injection abortions.

Chapter Notes

1. Hern, Warren. "Did I Violate the Partial-Birth Abortion Ban?" *Slate.* (2003)

   https://slate.com/technology/2003/10/did-i-violate-the-partial-birth-abortion-ban.html

2. "MedGyn Hern Ovum Evacuation Forceps" *MedGyn Products, Inc.*

   https://www.medgyn.com/product/hern-ovum-forceps/

3. Coolidge, Ardee. "The Reason Why So Many People Want to Eradicate Unborn Children with Down syndrome" *Care Net.* (2018)

   https://www.care-net.org/abundant-life-blog/the-reason-why-so-many-people-want-to-eradicate-unborn-children-with-down-syndrome

4. "Amniocentesis" *American Pregnancy Association.*

   https://americanpregnancy.org/prenatal-testing/amniocentesis/

5. Ibid

6. Daniilidis, A. et al. "A four-year retrospective study of amniocentesis: one centre experience." Hippokratia vol. 12,2 (2008): 113-5.

   https://www.ncbi.nlm.nih.gov/pmc/articles/PMC2464303/

7. Brown, David. "Late Term Abortions" *Washington Post.* (1996)

   https://www.washingtonpost.com/archive/lifestyle/wellness/1996/09/17/late-term-abortions/f15ae3a6-9711-45cc-9c13-e5160d293489/?noredirect=on

THE LETHAL INJECTION

Chapter 16

Lethal Injection, Partial Birth, Saline Infusion, and Third Trimester Abortions

---

To stop the heart, potassium chloride is administered directly after the vecuronium bromide. Without proper sedation, this stage would be extremely painful. The feeling has been likened to 'liquid fire' entering veins and snaking towards the heart.[1]

BBC Journalist Ben Bryant

This quote comes from a 2018 article about the execution of criminals in the United States. In this article, he describes a three drug regimen that is used for executions and ends with a fatal dose of potassium chloride. Potassium chloride stops the heart. In essence, the criminal dies of a heart attack, if he hasn't already been killed by one of the other drugs. Potassium chloride doesn't just stop the heart. It also causes violent muscle spasms, sending the criminal's body into convulsions.[2] This is why the criminal is given a drug to paralyze him.

Not every criminal prefers to be executed with potassium chloride. Some criminals are now asking to be executed with Digoxin. Digoxin is a heart medication used for heart failure and irregular heartbeats. But it is also being used off-label in physician-assisted suicide due to its affordability.[3] Off-label simply means that the drug is being prescribed for a purpose other than purpose for which the FDA approved the drug. Here again, an overdose of Digoxin causes the heart to stop. Not only is potassium chloride used in a cocktail of drugs, digoxin is also used as part of a cocktail in assisted suicides. The reason some criminals prefer to be executed with Digoxin is because it can be taken orally instead of with a lethal injection.[4]

Whether using potassium chloride to execute criminals or digoxin to commit suicide, additional drugs are used to try to make the death less violent and less painful. Drugs are used to render the person unconscious so that they don't suffer as much. And they are also used to paralyze people so that they don't thrash about violently.

Lethal drugs aren't just used on criminals and people who qualify legally for physician-assisted suicide. These lethal injections are also used in the

United States and around the world for third trimester abortions. Occasionally they are also used for second trimester abortions as well. These third trimester abortions are often called lethal injection abortions or heart attack abortions by those of us who are pro-life. It is hard to comprehend that people get third trimester abortions. Many people don't even understand that it is legal in some states to get third trimester abortions. Second and third trimester abortions are often referred to as "late-term abortions." Late-term abortion is a term widely used and understood among English speakers. It is not a technical medical term. Instead, it is plain language. The pro-choice movement hates this term because late-term abortions are overwhelmingly unpopular with the public. Instead they have come up with their own term, "later abortion." I'm not sure why the pro-choice movement thinks later abortion will be a more favourable term for their side. It seems likely to me that the pro-choice movement will have to dump this term as well in favour of something more ambiguous.

The reason it has become common to use the lethal injection in third trimester abortions is because these babies have a chance of survival outside the womb. In the third trimester they are past the generally considered point of viability at 24 weeks gestation. As we learned in chapter 10, their lungs have developed enough that they may be capable of breathing on their own. In the event that a woman wants an abortion of a viable or potentially viable baby, the abortionist wants to ensure that he is successful in killing it. Abortionists do not want to run the risk of an accidental live birth. A lethal injection while the child is still in the womb is the most efficient method devised by abortionists to ensure death. If the same doctor were to give the same lethal injection after birth, it would be considered murder. They have to get the job done before birth. Abortionists typically use digoxin because of its affordability, but some also use potassium chloride. Unlike criminals, however, no other drugs are used to make the death of the child less violent. No drugs are used to render the child unconscious or paralyzed. Even many on the pro-choice side will admit that the child is conscious and able to feel pain after 24-27 weeks gestation.

Pro-choice advocates will often claim that these abortions are rare as justification for third trimester abortions. They point to the fact that these abortions make up just over 1% of all abortions.[5] In my opinion, the fact that these abortions exist at all is unconscionable. I don't consider "rare" to be a justification. Instead, "rare" is an admission that these abortions do occur. So how many of these third trimester abortions are being performed? The best data available is from the CDC's last abortion surveillance report from 2015. They reported a total of 5,597 abortions at or after 21 weeks gestation in 2015, making up 1.3% of total abortions. Twenty one weeks of gestation isn't considered viable but is the point when some micro preemies have a chance of survival after birth. Even at this "rare" percentage, the number is staggering. Delaware reported seven of these abortions for .3% of all Delaware abortions. The second highest state that reported was Colorado with 298 third trimester abortions at 3% of total abortions. Colorado was higher because that is where Warren Hern specializes in late-term abortions. The highest state was New Mexico with 336 third trimester abortions at 7% of total abortions. New Mexico also has a facility that specializes in late-term abortions.[6]

The real number of third trimester abortions, however, is far higher than the 5,597 reported. Eleven states and the District of Columbia did not report their numbers. States that do not report tend to be much more permissive of late-term abortions. These include California, New York, and Maryland, which are home to late-term abortion facilities. This means that the real number of late-term abortions are sure to be much higher than reported to the CDC. The percentage of total abortions is surely higher than 1.3%. And the total number of these abortions is likely between 10,000-

15,000. These numbers are likely to rise as a number of liberal states have expanded access to third trimester abortions after the election of President Trump.

The state of New York shocked the nation when they legalized third trimester abortions and Governor Cuomo ordered the One World Trade Center to be lit in pink to celebrate. Delaware was the first state to expand late-term abortions in the wake of the election of President Trump when it passed Senate Bill 5 in 2017. It's an unfortunate distinction to be labelled not only "the first state" but also the first state in late-term abortions! Delaware didn't get much notice because we are a small state but also because the governor signed Senate Bill 5 behind closed doors and without any celebration. New York, on the other hand, celebrated loudly with cheering at the signing ceremony and by lighting the One World Trade Center in pink. The celebration and brazen support of third trimester abortions was the wakeup call that the pro-life movement needed.

The pro-choice movement also justifies these late-term abortions by claiming that nearly all of them are to protect the mother's health or because the fetus has a severe or fatal birth defect or medical condition. They never cite statistics or proof that this is why people have late-term abortions because no such proof exists. The best they can do is to give some examples of women who have had abortions for those reasons. What little data we have suggests that this is not true. Florida, for example, tracks the reasons that people claim they are having abortions. They track by trimester. There were only two abortions in the third trimester reported in 2018. That is not enough to make any kind of statistical claims. But 4,268 second trimester abortions were reported. Only 10.4% were because of a fetal medical condition or disability. And only .7% were because the mother's life was in danger.[7]

A study published in 2013 in the journal *Perspectives on Sexual and Reproductive Health* looked at the reasons women wait till after 20 weeks to get an abortion. Fetal medical conditions and the health of the mother were so insignificant as to not even show up in their research. The authors wrote, "But data suggest that most women seeking later terminations are not doing so for reasons of fetal anomaly or life endangerment."[8]

Further, we can look at those doctors who specialize in late-term abortion to see what they say are the reasons women get late-term abortions. Occasionally, they are honest enough to admit that many of these late-term abortions are elective abortions. Martin Haskell is one of the few doctors to do abortions in the third trimester. In the early 90s, he was quoted in a medical publication as saying, "I'll be quite frank: most of my abortions are elective in that

20-24 week range. In my particular case, probably 20 percent are for genetic reasons. And the other 80 percent are purely elective."[9]

Lethal injection abortions are a much more lengthy and costly process than most other abortions. The woman first has to find an abortion clinic that does third trimester abortions. There are only a handful of clinics around the country that specialize in these risky procedures. She has to go to the clinic for an evaluation and consultation. The abortionist has to examine her, which includes an ultrasound to attempt to determine how large and how old the fetus is. The abortionist then has to discuss which options for the abortion he is willing to do. There are different variations of lethal injection abortions. Once it is determined what variation of abortion they are doing, she has to sign the consent forms consenting to the abortion. She also has to have the money to pay for it. Third trimester abortions cost thousands of dollars. If she has the money, they can start the abortion procedure.

The abortion begins with the lethal injection. The abortionist will have a large needle to inject the fetus with his drug of choice. This abortion requires that the needle be guided by an ultrasound. The needle can either be injected through the abdomen or transvaginally. If it is done through the abdomen, the abortionist simply pushes the needle through the

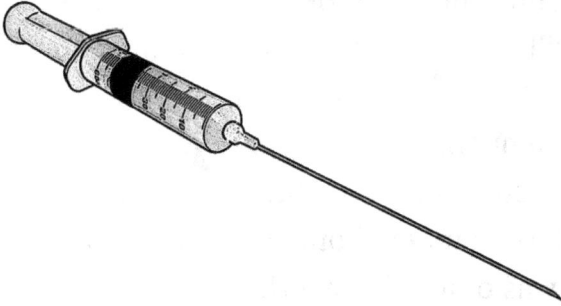

Needle Used for Lethal Injection Abortions

woman's belly till he reaches the fetus to inject it with the drug of choice. Transvaginally means that the abortionist inserts the needle through the vagina and the cervix to reach the fetus. The goal is to get the needle into the body of the fetus to inject the fatal drug. Even if the needle misses and it is injected into the amniotic fluid, it can still successfully kill the child. This point in the abortion can be a particularly traumatic time as the abortionist and the mother are likely to feel the baby struggle as it is injected. At this visit the abortionist will also insert laminaria to begin dilating the cervix. Laminaria is a seaweed that absorbs fluid and expands to dilate the cervix.

This second appointment typically occurs the next day. This appointment begins with an ultrasound to confirm that the fetus is actually dead. Sometimes if the abortionist fails to inject the fetus, it may survive. Typically if the abortionist misses the fetus and injects the drug into the amniotic fluid, it will still kill the fetus. But it will also take significantly longer to kill it. And occasionally, the doctor discovers at the second appointment that the fetus survived. In this case, the lethal injection must be administered a second time. You will remember from chapter 4 that Dr. Kermit Gosnell was doing these lethal injection abortions, but was not confirming that they were dead. Instead, if the baby happened to be alive after it was born, he would kill it after birth. You will also remember from chapter 4 that these abortions were done across state lines in order to avoid being charged with illegal late-term abortions.

Assuming the fetus was successfully killed, the next step is to remove it. This is done typically in one of two ways: through induced labor or through dismemberment. Induced labor simply means that labor is induced and the woman births an intact and dead baby. Dismemberment means that the abortion is finished with a forceps. The abortionist removes the fetus piece by piece, the same as a second trimester forceps abortion. Some abortionists believe that the bones soften by the next day after the fetus died, making it easier to dismember.[10] Sometimes the abortionist may give the mother the option of which method to do. Forceps abortions in the third trimester are at high risk for uterine perforation due to the large size of the fetus and the size of the uterus which has been stretched to accommodate the child.

Lethal injection abortions are often confused with partial birth abortions, which are done in the second and third trimester. I frequently get questions during my Fetal Beauty presentations about partial birth abortions. Lethal injection abortions are not the same as partial birth abortions. Partial birth abortions are a crime under federal law and also under many state laws. A partial birth abortion is when a living fetus is delivered all but the head. A suction tube is inserted into the back of the head. The brain is removed causing the head to collapse. The primary reason partial birth abortions were done was to solve the problem of the head. The head is a challenge to abortionists because it is large and round. By collapsing the head, it makes the fetus easier to deliver and doesn't result in sharp skull fragments inside the uterus. The textbook *Management of Unintended and Abnormal Pregnancy* describes the head problem stating, "Because the cranium represents the largest and least compressible structure, it often requires decompression."[11]

There are many other variations of the procedure that fall within the federal legal definition of partial birth abortion, one of which causes the death of the fetus by decapitation. A scissors or forceps is used to separate the head from the body while the head is still in the birth canal.[12] Congress made partial birth abortion a crime in 2003 and the law was upheld by the Supreme Court in 2007. However, an abortion is not considered a partial birth abortion if the fetus is already dead before entering the birth canal. This is where lethal injection abortions come in. A lethal injection can cause the death of the fetus before partially delivering the child. There are other tricks that abortionists use as well. An abortionist may remove the arms and legs before doing the partial birth abortion. By removing limbs, it is impossible to prove whether or not the fetus was alive when it was partially delivered. Another method is to cut the umbilical cord first to cause the death of the fetus before partially delivering it. The generally accepted alternative, however, has been lethal injection abortions. According to the Society of Family Planning, lethal injection abortions are done widely "to avoid signs of life at delivery."[13]

Another late-term abortion procedure often confused with lethal injection and partial birth abortion is the saline abortion. This type of abortion is also called an infusion or instillation abortion and can use saline or other toxic substances to kill the fetus. This abortion procedure is rare in the United Sates today but was much more common in the 1970s. Many of the survivors of failed abortions that we hear about today were survivors of saline abortions. In this abortion, some of the amniotic fluid is removed to make room for the toxic substance. The saline solution or other toxin is injected into the amniotic fluid. A healthy fetus practices breathing in the womb. Instead of breathing air, she breathes amniotic fluid. But in a saline abortion, she ingests the toxin. Saline solution has a burning effect both on the inside and outside of the fetus. The result is a horribly disfigured body. After the death of the fetus, the woman goes into labor and delivers the dead fetus. Sometimes lethal injection abortions accidently become infusion abortions. If the doctor misses the body of the fetus with the injection, he may then accidentally inject the drug into the amniotic fluid. The drug still kills the child, but it is killed through infusion into the fluid rather than injection into the body. As I'm sure you can see at this point, there are many variations in the ways that doctors do abortions. Abortions don't fit neatly into categories. What they all have in common is that they are all violent acts.

Third trimester abortions are the most violent abortions and also the abortions the public overwhelmingly wants to prohibit legally. This is true due to the fact that the larger and more developed the fetus is, the more violent the abortion must be to successfully kill and remove it. We know that all abortions are violent in so far as all are acts of killing a living human being. But the amount of bodily injuries done increases as the child grows. By the time we get into the third trimester, it is hard to judge which abortions are more violent than others. For example, is an abortion that involves crushing of the head with a forceps more or less violent than one that involves aspirating the brain matter? I would say that crushing with the forceps is more violent as it destroys the outside of the head and skull more than just aspirating the brain. But that's just my subjective opinion. I think what most people can agree with is that the lethal injection is less violent than other forms of second and third trimester abortions. A drug used to stop the heart seems to be a less violent way to kill the fetus because the amount of injuries done to the body are less than dismemberment with a forceps, aspiration of the brain, crushing of the head, or decapitation. This is the motivation behind several recent state laws that prohibit the killing of the fetus through dismemberment. The goal of these laws is to force abortionists to use lethal injection to kill the fetus in the same way that the partial birth abortion ban has forced abortionist to resort to the lethal injection. The motivation is to force abortion doctors to use a less violent method of killing the fetus.

While the lethal injection seems less violent than other late-term abortions, it is still a more violent procedure than the lethal injection used on criminals. No drugs are used to render the fetus unconscious or paralysed. As a result, the fetus is likely to thrash about violently. The amount of thrashing is likely to increase further into pregnancy as the fetus grows larger and stronger. Also, the lethal injection doesn't stop further violence after death. While injuries to the body after death wouldn't be considered violence in the same way as injuries before death, those injuries are still significant. It is still a human body that is being destroyed. In our culture, the human body, even though death has occurred, it still worthy of a certain amount of dignity and respect. This is why states have laws against the desecration of bodies. This is also why the State of Indiana passed a law requiring the burial or cremation of fetal bodies after abortion. That law was upheld by the U.S. Supreme Court in 2019 in Box v. Planned Parenthood of Indiana and Kentucky Inc.

Another aspect of late-term abortions that is widely misunderstood is the fact that sometimes babies are accidently born alive after botched late-term abortions. Also misunderstood is what is done with those babies after they are born. Many on the pro-choice side refuse to admit that this occurs. They don't want to admit that babies can survive failed abortions, especially viable babies. But we know that this occurs because these babies have grown up to be adults and those adults are speaking out. A few of the more well-known survivors of botched abortions include Gianna Jessen, Melissa Ohden, Claire Culwell, and Josiah Presley. I would encourage you to learn more about each of these survivors. Each survivor has an incredible story of how they beat the odds and are alive today to talk about it.

There are a large number of ways that people can survive abortion attempts. They can be loosely categorized into two groups: those that survived failed abortions before viability and went on to develop to full term and those that were born alive during a third trimester abortion gone wrong. The first group can include situations where the fetus survives the abortion pill, where the abortionist fails to kill and remove the fetus, or where the abortionist fails to detect the presence of another fetus when there are twins or multiples. When an abortion fails, the pregnant mother may decide not to go back for a second abortion to finish the job. One of the more famous examples is that of Claire Culwell who survived when her mother went for a surgical abortion. The abortionist got her twin but missed her.[14] You can visit www.ClaireCulwell.com to learn more about Claire's story.

But for this chapter we are more concerned about babies that are accidently born alive during a third trimester procedure or at the end of the second trimester. These are babies that may have a chance of survival if given the best level of care in a NICU. There are three things that can happen to a baby that is born alive: the doctor can send it to a NICU and attempt to save it, the doctor can refuse care until it dies, or the doctor can actively kill it. In 2002, Congress passed and President George W. Bush signed into law the Born Alive Infants Protection Act which declared that babies born alive after a failed abortion should be treated equally under the law.[15] Under this law, doctors are obligated to treat these babies the same as any other premature baby. But under the Born Alive Infants Protection Act, there are no teeth to punish abortion doctors who refuse to give these babies born alive after a failed abortion an equal chance at life. In 2019, Republicans in Congress made numerous attempts to add teeth to this law, but failed as pro-choice Democrats opposed these efforts to strengthen the Born Alive Infants Protection Act. I'm not pointing this out for partisan reasons as I am myself an active member of the Democrat party. I do hope that my own party will come to understand that true progress and true equality will move us away from abortion violence.

Without any enforcement of the Born Alive Infants Protection Act, if a baby isn't given NICU care, it may simply be refused care until it dies. Unfortunately, this seems to be the method of choice for many abortionists.

Even if he violates the Born Alive Infants Protection Act, he won't be prosecuted or punished. Babies at 24 weeks gestation have a greater than 50% chance of survival with proper NICU care. But without a NICU, they cannot survive on their own until several weeks later. And so babies that are viable and very likely to survive can be killed by simply denying care. Many individuals have given their testimonies about witnessing babies that were denied care. One such woman is a nurse who I know personally. Leslie Dean wrote:

> *The doctor estimated him to be between 19-20 weeks. His body had been badly burned, and the expression on his face was unmistakably one of intense pain. He was still alive.*
>
> *The doctor explained if the eyes were not "fixed" we may need to resuscitate. As he held up the baby to check the eyes, the mother saw him and began to scream uncontrollably: "Oh, God, what have I done?"*
>
> *Declaring the eyes were fixed, he dropped the baby in a bucket on the floor where I saw it moving and gasping for breath, and then died.*[16]

The third option is to actively kill the child outright. The late-term abortionist Kermit Gosnell is spending life in prison for actively killing babies after they were born alive. Actively killing after birth is not as common due to the risk of being prosecuted and jailed for murder. But there have been testimonies by witnesses of other doctors doing similar killings after babies were born alive. A doctor might decide to actively kill the child if it could survive without the care of a NICU or if it is necessary to harvest fresh organs.

The pro-choice movement is currently very opposed to the idea of making abortion doctors save babies that are born alive. They now frequently claim that babies aren't born alive, that late-term abortions are not real, and that babies aren't being denied care. At the time of this writing, these types of claims have been made by a number of the Democrat candidates for President. Unfortunately, there are very few statistics about how many babies are born alive after failed abortions and what happens to those babies. There are some reports out of Canada, Australia, and the United Kingdom. But those reports aren't thorough enough and reliable enough to give us a good picture of how many of these babies exist. We do know that babies are born alive after failed abortions because of those few reports and because of the stories told by witnesses. But we also know that babies are born alive after failed abortions because the abortion industry on rare instances admits to it. In *Management of Unintended and Abnormal Pregnancy* the authors state that one of the reasons abortionists might use the lethal injection is because "they desire to avoid the possibility of unscheduled delivery of a live fetus."[17]

This has been one of the longest and most difficult chapters to write. I can't tell you how much I appreciate you putting yourself through the emotional pain of reading this chapter. There were many mornings when I wrote only a few sentences before I had to put it down and walk away. But it's incredibly important that everyone understand what our society has allowed to happen since Roe v. Wade and Doe v. Bolton. And it is equally important that every one of us stand up to the pro-choice movement and put an end to this. This is happening on our watch.

Chapter Notes

1. Bryant, Ben. "Life and Death Row: How the lethal injection kills" BBC. (2018)

    https://www.bbc.co.uk/bbcthree/article/cd49a818-5645-4a94-832e-d22860804779

2. Ibid

3. Aleccia, JoNel. "Northwest doctors rethink aid-in-dying drugs to avoid prolonged deaths" *The Seattle Times*. (2017)

    https://www.seattletimes.com/seattle-news/health/northwest-doctors-rethink-aid-in-dying-drugs-to-avoid-prolonged-deaths/

4. Harcourt, Bernard. "Second Amended Complaint." Hamm v. Dunn. Filed in the United States District Court for the Northern District of Alabama. (2018)

    http://blogs.law.columbia.edu/update-hamm-v-alabama/files/2018/03/94-1-Hamm-Second-   Amended-Complaint-STAMPED.pdf

5. "Later Abortions" *Guttmacher Institute*. (2017)

    https://www.guttmacher.org/evidence-you-can-use/later-abortion

6. Jatlaoui, T. et al. "Abortion Surveillance – United States, 2015" *Centers for Disease Control and Prevention*. (2018)

    http://dx.doi.org/10.15585/mmwr.ss6713a1

7. "Reported Induced Terminations of Pregnancy (ITOP) by Reason, by Trimester" *Agency for Health Care Administration*. (2019)

    https://ahca.myflorida.com/MCHQ/Central_Services/Training_Support/docs/TrimesterBy Reason_2018.pdf

8. Foster, Diana and Katrina Kimport. "Who Seeks Abortions at or After 20 Weeks?" *Wiley Online Library*. (2013)

    https://doi.org/10.1363/4521013

9. Brown, David. "Late Term Abortions" Washington Post. (1996)

    https://www.washingtonpost.com/archive/lifestyle/ wellness/1996/09/17/late-term-abortions/f15ae3a6-9711- 45cc-9c13-e5160d293489/?noredirect=on&utm_ term=.4fba3ce7b6b1

10. Paul, Maureen, et al. *Management of Unintended and Abnormal Pregnancy* Hoboken: Wiley-Blackwell, 2009, pp. 166

11. Paul, pp. 173

12. Gorney, Cynthia. "Gambling with abortion: why both sides think they have everything to lose." The Free Library. (2004)

    https://www.thefreelibrary.com/Gambling+with+abortion% 3A+why+both+sides+think+they+have+everything+to...-a0126 194929

    See exchange between DOJ lawyer Quinlivan and Physician "Doe." This exchange describes decapitation with a scissors. I've also seen video of decapitation with a forceps.

13. Diedrich, Justin and Eleanor Drey. "Clinical Guidelines: Induction of fetal demise before abortion" *Society of Family Planning.* (2010)

    https://www.societyfp.org/_documents/resources/ Induction ofFetalDemise.pdf

14. Culwell, Claire. "My Story"

    http://www.claireculwell.com/my-story.html

15. H.R. 2175. "An act to protect infants who are born alive." 107th Congress, 2D Session. (2002)

    http://www.nrlc.org/uploads/bornaliveinfants/Baipatext. pdf

16. Dean, Leslie. "I've Had 2 Abortions. Here's Why I Support Alabama's Pro-Life Law." *The Daily Signal.* (2019)

    https://www.dailysignal.com/2019/05/19/ive-had-2- abortions-heres-why-i-support-alabamas-pro-life-law/

17. Paul, pp. 169

POC

Chapter 17

Abortion Industry Propaganda (Part A) Deceptive Euphemisms & Confusion about When Life Begins

> *[Menstrual extraction] originated as a euphemism for early abortion prior to the legalization of abortion and was perceived by its originators as a useful deception.*[1]
>
> Late-Term Abortion Specialist Dr. Warren Hern

I was recently moved by the story of a man named Kevin in an email update from the pro-life organization 40 Days for Life. Kevin used to work in pathology where they would examine organs and tissues. One day Kevin received in his lab an unborn baby girl that had been aborted. But the worst part for Kevin wasn't seeing the corpse. It was the way the corpse was labeled. On the label of the package was "Product of conception (POC)." This tiny girl didn't even have the respect of being recognized as human. Kevin was so disturbed by the blatant propaganda on that package that he joined the pro-life movement and became a pro-life pregnancy care center director.[2]

Kevin's story is an example of the primary form of propaganda used by the abortion industry, deceptive and medically inaccurate euphemisms. In the next two chapters we will look at four types of propaganda most often used: deceptive euphemisms, confusion about when life begins, deceptive illustrations and pictures, and pregnancy alarmism. When I talk about propaganda in this book, I am referring to deliberately using false, misleading, or deceptive information and claims. That's in contrast to good faith pro-choice arguments. Propaganda is not created and introduced in good faith. The purpose of this book is not necessarily to answer pro-choice arguments. There are plenty of other books that do. In these two chapters we are looking at some of the deliberate ways that false and misleading information is used to deceive the public into supporting abortion.

When I refer to the abortion industry, I am referring to Planned Parenthood and the various independent abortion clinics that make up the industry. I am not referring to the pro-choice movement as a whole. The abortion industry operates the same as any other industry. There are independent clinics that are incorporated as for-profit businesses. And then there is Planned Parenthood which is incorporated as a tax-exempt organization. But don't let the tax-exemption fool you. Many large organizations across the country are tax-exempt. Many of them are also very wealthy and pay their executives very well. This includes many wealthy hospitals, trade organizations, and sports leagues. Tax-exempt organizations like Planned Parenthood are profitable members of their respective industries. Planned Parenthood's annual report reveals that it is primarily an abortion provider and that it is enormously wealthy. This organization continues to increase its market share and its profits. In the fiscal year ending in 2018, its total assets increased from 1.6 billion to 1.9 billion dollars.[3]

Planned Parenthood and the independent clinics together make up the abortion industry. They have a trade association called the National Abortion Federation which acts in the same way that any other industry trade group would. It is not unusual for large industries to band together to protect their interests and their profits. In the same way that realtors have the National Association of Realtors and lawyers have the American Bar Association, abortion clinics have the National Abortion Federation. It's also not unusual for industries to spend money on politics in order to protect their interests and profits. The abortion industry spends tens of millions each election cycle on political candidates that will expand abortion. Since demand for abortions is in decline, the abortion industry is relying on pro-choice politicians to expand abortion into the third trimester, repeal regulations, and fund abortions with tax dollars. Some pro-choice people are uncomfortable admitting that there is an abortion industry that is motivated by profit. But it shouldn't surprise anyone that there is. The abortion industry is behaving the way we would expect any other industry to behave. What makes abortion different than most industries is years of conditioning of the work culture of the industry to rely on propaganda to keep abortion legal. In the years leading up to Roe, they learned that propaganda works. Because it is useful, they have been conditioned to rely heavily on it.

The first and most widely used form of propaganda is deceptive and medically inaccurate euphemisms. The abortion industry does not use euphemisms the way that most people do. For example, if your pet dog or cat is sick and suffering, you might put your pet to sleep. When I say "to sleep" I mean that you would kill it, not that you might put it to bed for the night. Everyone knows what I really mean. Euphemisms are words or phrases used to say something uncomfortable or embarrassing in a more mild or polite way. The purpose of a euphemism is not to confuse or deceive people. When I say that I am putting my dog to sleep, I am not trying to convince people that my dog is tired and needs to go to bed. Everyone knows that I am talking about killing my dog.

This is not how the abortion industry uses euphemisms. They use euphemisms specifically to confuse and deceive. And so if people don't understand what you are saying, then it isn't a true euphemism. It's just a lie. It would be as if I were actually trying to convince people that my dog was tired and I was putting it to bed because I don't want them to know that I am actually killing it. The abortion industry doesn't want you to know that it is actually killing tiny humans.

The abortion industry is flush with deceptive euphemisms. They are so common that most people don't even notice them.[4] Even the word abortion is a type of euphemism. They don't want to admit that they are killing, so they use words like termination and abortion instead. They don't want to say that they are killing a tiny human, so instead they say they are aborting a pregnancy. But what happens when everyone begins to realize that abortion actually means killing? When most people are no longer deceived and realize what the euphemism actually means, then the abortion industry must come up with a new euphemism. One such new euphemism is the word management. They aren't killing. They aren't aborting. They are managing. Managing sounds like it has to be a good thing, right?

One example of these euphemisms comes from a hospital network in my area called Christiana Care. On their campus is a section called the "Center for Reproductive Health." If you look at their list of services, it conveniently doesn't mention abortion. Very few people realize that this is, among other things, an abortion clinic. And it's not just an abortion clinic. It is a clinic that targets disabled babies later in the pregnancy. How do we know that? All we have to do is follow the euphemisms. Their website lists as one of its services "Management of miscarriage and fetal anomaly."[5]

This description is deceptive on so many levels. They don't want to admit to doing abortions and so they take abortions and miscarriages, which are not abortions, and lump them both under the word management. So we see that management isn't just a word for removing the baby after it has died, a miscarriage. It is also a word for aborting the baby in the case of "fetal anomaly." But what is fetal anomaly? This is another euphemism, this time to describe babies with various medical conditions and disabilities. This euphemism can include anything as serious as the baby having a life threatening medical condition to as minor as cleft palate. In Delaware, as is the case in much of the country, abortion is legal up till birth if the baby has a life threatening "fetal anomaly."

One of the more common conditions resulting in "management of fetal anomaly" is Down Syndrome. Down Syndrome, of course, does not mean that death is imminent. Many people with Down Syndrome go on to live happy fulfilling lives. Most of these babies, however, are aborted. Because most of these babies aren't diagnosed with Down Syndrome until the middle to end of the second trimester, they are subjected to more horrific late-term abortions.

And so when Christiana Care or your local hospital says that they do "management of fetal anomalies," understand that what they are really doing is late-term abortions on babies with disabilities.

Another euphemism which is used to replace abortion is "healthcare." This is the latest euphemism being pushed by Planned Parenthood. They frequently have signs at their abortion clinics saying "Healthcare happens here." Planned Parenthood's leaders and politicians affiliated with them frequently refer to healthcare instead of abortion. This is one of the reasons we know that the pro-life side is winning. We have been so successful that the pro-choice side is abandoning the word abortion. They can't use the word abortion because too many people now know what it really is. This should be a lesson to the pro-life movement. We can reclaim some of these euphemisms and render them useless as propaganda.

Planned Parenthood isn't just adopting the euphemism healthcare to avoid the word abortion. They are doing it to try to create political support for abortion. Most people don't like abortion. Most people know in their hearts that abortion is horrifically violent. But healthcare sounds like a great thing. We all care about our health. And we all like to be cared for. But in reality, it is deceptive at its core. Abortion isn't healthcare. Healthcare means to care for someone's health. But whose health is abortion caring for? It isn't caring for the fetus' health. The fetus ends up dead in every successful abortion. The goal of the abortion is for it to be dead and removed.

But isn't abortion caring for the mother's health? No, it isn't actually caring for the mother's health. It is true that in rare cases the mother may face a life threatening condition before the baby is viable. In a case like this, you may have to treat the mother, even if it means the death of the fetus because the alternative is losing both. The most common example is a surgery to remove an embryo or fetus that has implanted in the fallopian tube. This kind of ectopic pregnancy will certainly kill both mother and child. The surgeon actually removes the fetus intact from the mother's fallopian tube. That tiny child continues to live for a time after being removed but eventually dies from a lack of oxygen and nutrition no longer supplied by its mother. This is certainly healthcare. It is saving the life of the mother. And this surgery does indirectly result in the death of the fetus. But this is not what we are talking about when we debate abortion. This surgery is never done in an abortion clinic or at Planned Parenthood. It is only an abortion in a broad sense of the word. The abortions done at abortion clinics like Planned Parenthood are nearly always elective abortions. They are elective because they aren't being done for legitimate health reasons. Instead, they are being done for personal reasons, such as the fear of dropping out of school or the fear of being poor. This isn't healthcare. Elective abortions do nothing to care for the mother's health. And yet Planned Parenthood insists that "healthcare happens here" in order to build political support.

On rare occasions, those in the abortion field are honest about and even push back against deceptive euphemisms. One such euphemism that is controversial in the abortion industry is "menstrual extraction." Menstrual extraction is a deceptive euphemism for a very early vacuum abortion when pregnancy has not been confirmed. It is deceptive because you can't extract a woman's period. If the woman is not pregnant, then the abortionist is only removing the uterine lining. If the woman is pregnant, then it is an abortion. In menstrual extraction, the abortionist does not examine the removed tissue lest he see the tiny body parts and learn positively that he performed an abortion. Before

Roe, when early elective abortions were illegal in most states, an abortionist could claim that he was not performing abortions since the woman wasn't diagnosed as pregnant. And so this was a way to perform illegal abortions and avoid prosecution. When pro-choice activists talk about illegal abortions, they try to conjure up images of a woman mutilating herself with a coat hanger. While it's true that some women did and still do bizarre self-administered abortions, menstrual extraction is far more representative of illegal abortions than a coat hanger.

Abortionist Dr. Warren Hern is one abortionist that takes issue with the term menstrual extraction. He states in his book *Abortion Practice*, "The term originated as a euphemism for early abortion prior to the legalization of abortion and was perceived by its originators as a useful deception."[6] He goes on to write, "It is an inaccurate term, since the menstrual period is not extracted. The woman has either an endometrial aspiration (if she is not pregnant) or an early abortion (if she is pregnant)."[7] So why then would abortionists like Hern oppose this deceptive euphemism while embracing so many others? Notice that Hern says proponents of menstrual extraction saw it as a useful deception. The idea is that deception is justified so long as it is useful. Hern was opposed to the term and the practice because it subjected women who were not pregnant with the unnecessary risk of complications.

In other words, deception is only wrong in his view when it leads to practices that have higher complication rates. Therefore, most deceptive euphemisms are good because they are useful in accomplishing the industry's goals. Most of the controversy over menstrual extraction disappeared after Roe. Some modern feminists, however, are romanticizing and reviving the practice of menstrual extraction and dangerous black market abortions.[8]

Another form of propaganda just as common as deceptive euphemisms is the deliberate sowing of confusion about when life begins. We looked at this question in detail in chapter 6. The main points to remember from that chapter are that biologists are the experts on living things. Biology shows conclusively that a new living human begins when the 46 chromosomes finish fusing together during conception, forming a complete set of new human chromosomes and starting the growth and development of a new living human. The pro-choice movement needs to keep the public distracted and confused in order to avoid the real issue: whether the violence of abortion is justified. They can't allow abortion to be an honest discussion about the violence because they know they will lose public opinion. And so sowing needless confusion about when life begins is a vital piece of propaganda for the pro-choice movement.

Chapter Notes:

1. Hern, Warren. *Abortion Practice*, Philadelphia: J.B. Lippincott Company, 1984, pp 120.

2. Carney, Shawn. "Allow New York to be your answer" 40 *Days For Life.* (2019)

   https://www.40daysforlife.com/2019/01/29/allow-new-york-to-be-your-answer/

3. Desanctis, Alexandra. "Planned Parenthood's New Annual Report Disproves Its Own Narrative" *National Review.* (2019)

   https://www.nationalreview.com/2019/01/planned-parenthood -annual-report-disproves-narrative/

4. Paul, Maureen, et al. *Management of Unintended and Abnormal Pregnancy* Hoboken: Wiley-Blackwell, 2009, pp. 135-166

   A review of chapter 10 found twenty-two different euphemisms for fetus and embryo.

5. "Family Planning & Maternity" *Christiana Care Health System.*

   https://christianacare.org/services/women/family-planning/maternity/pregnancy-testing/

6. Hern, pp. 120

7. Hern, pp. 121

8. Block, Jennifer. "Not Your Grandmother's Illegal Abortion" *The Cut.* (2019)

   https://www.thecut.com/2019/07/excerpt-from-everything-below-the-waist.html

A FORMLESS BLOB

Chapter 18

Abortion Industry Propaganda (Part B) Deceptive Illustrations & Pregnancy Alarmism

The third category of pro-choice propaganda is deceptive and inaccurate illustrations and pictures. This category of propaganda is rarely seen outside of the abortion industry itself. In fact, I had never seen these pictures and illustrations until I bought the abortion textbook *Management of Unintended and Abnormal Pregnancy*. And so the target audience of this particular propaganda appears to be people in the medical field or in medical school, abortion clinic staff, and potential abortion clinic staff. It's unlikely that anyone outside of the medical field would stumble upon this book and publications like it. This book is published by the National Abortion Federation (NAF), the official abortion trade group that describes itself as "the professional association of abortion providers in the USA and Candada."[1]

The most widespread example of deceptive illustrations is a set of medical illustrations created by medical illustrator and pro-choice activist Lisa Penalver and used throughout NAF publications and abortion training manuals around the world. Penalver's illustrations depict the embryo or fetus as a formless blob of tissue. This is a large set of illustrations commissioned by the NAF illustrating various aspects of abortion in greyscale and in color. Some of these illustrations are quite impressive in their coloring and detail. But what they all have in common is the distinctive depiction of the embryo or fetus as an oval-shaped blob.[2] What is especially egregious about these deceptive and dehumanizing illustrations is that they are created by medical professionals for medical professionals. As you know from chapters 8 and 10 of this book, the human is a highly complex creature from the moment of conception. Even to depict the embryo as a formless blob of tissue in the first few days of its existence is a stretch. It certainly isn't an oval-shaped, formless blob when a woman goes for an abortion. But this is the great lengths that the abortion industry will go through to convince themselves that what they are killing is less than human.

Another such deceptive illustration, also by Penalver, is a second trimester forceps abortion in which the uterus is empty. This set of two illustrations is for the purpose of showing how to hold and operate the forceps in the uterus. The only problem is that there is no fetus in this illustration. It is an illustration of an abortion on a woman who is not pregnant. The caption for the illustrations describes how to use the jaws of the forceps to grasp the fetal parts. Only there are no fetal parts in the illustration![3] To show the fetal parts would be to recognize the highly-developed body of a prenatal human child.

Another deceptive, commonly used tool are pictures of the "products of conception" that conveniently leave out the embryo or fetus.[4] These pictures are numerous throughout the abortion industry. In each one the embryo or fetus has been carefully removed or destroyed beyond recognition, leaving only the supporting structures as recognizable so as not to show the human that has been killed. These are pictures taken in the first trimester. The abortionists are particularly fond of pointing out the gestational sac that once held the embryo and the villi, which is early placenta tissue. Some of these pictures are taken later in the first trimester when it is a fetus and looks clearly like a baby. And yet the fetus is conveniently excluded from the picture.

The closest this particular textbook ever gets to revealing the humanity and form of the child is with the use of a few old black and white ultrasound pictures. These ultrasounds are not the nice, crisp ultrasounds that many pregnant mothers are used to seeing today. The technology in modern ultrasound machines are almost too good to be believed. With our first daughter, the ultrasound machine was so advanced and the technician was so skilled that she was able to accurately tell the sex of our baby at only about 12 weeks gestation. That is how far we've come. But the pictures used in this textbook look like they are decades old. They are so grainy that you can barely see the shape of the fetus. At best you can see the round shape of a head and the rough outline of the body. But you can't see enough to see what a fetus actually looks like. These deceptive and inaccurate illustrations and pictures show how far the abortion industry is willing to go to cover up the violence and to deceive, if possible, even themselves.

The final type of propaganda we will look at is something that I call pregnancy alarmism. Alarmism is an exaggerated belief that something is dangerous so as to cause needless fear and panic. Pregnancy alarmism is the belief that every pregnancy is dangerous and scary. This propaganda is aimed in particular at young single women. The goal is to convince young women and the people who love them that they need abortion because pregnancy is deadly and scary. They see every abortion as necessary to save the life of the woman. This is why you see pro-choice people claiming that abortion saves lives. In their minds, they are heroes, literally saving women from death. And those of us who are pro-life are waging a "war on women." Pregnancy alarmism is an incredibly powerful and false ideology used to whip people up into a frenzy.

Abortion doctor Warren Hern recently wrote an opinion piece in the *New York Times* peddling pregnancy alarmism. His article was titled "Pregnancy Kills. Abortion Saves Lives." The subtitle was "Every pregnancy poses a serious health risk to the mother." (It is an interesting side note that Hearn refers to every pregnant woman as a "mother.") Hern goes on to make the following alarmist claims in his essay:

> *Pregnancy is a life threatening condition. Women die from being pregnant.*
>
> *Pregnancy is dangerous: abortion can be life saving.*
>
> *A woman's life and health are at risk from the moment that a pregnancy exists in her body, whether she wants to be pregnant or not.*[5]

The evidence that the abortion industry uses and that Hern uses in his essay are statistics showing a higher mortality rate for pregnant woman than for women who get abortions. The argument is simply that abortion is safer, therefore, saving lives. The numbers Hern uses are 20.7 deaths per 100,000 live births for women who do not get abortions and .7 deaths per 100,000 abortions. The latest maternal mortality rate from the CDC is 17.2 per 100,000 live births.[6]

There is no doubt in my mind that we have a maternal mortality crisis in America. Our maternal mortality rate is on the rise, which means that we are tracking an increasing number of women dying in connection to pregnancy and childbirth. Our maternal mortality rate is higher than much of the developed world and the maternal mortality rate for African-American women is substantially higher than for Caucasian women. Some conservatives may want to argue with me a little bit about these claims, but I accept them. And I think the pro-life movement should be the first ones to call this a maternal mortality crisis and to make it a priority.

It's also important to note that not only are African-American women more likely to die from pregnancy and childbirth, they are also more likely to die from complications from abortion. A study from a pro-choice researcher found that black women are 2.75 times more likely to die from an abortion than white women.[7]

Once you understand that abortion is a violent act, however, the idea that abortion is a solution to the maternal mortality crisis is absurd. The abortion industry isn't proposing other solutions to this crisis. They aren't advocating for better healthcare for pregnant women. They aren't arguing for more preventative healthcare for pregnant women. They aren't arguing for better research on maternal mortality. Instead they are arguing that nearly a million violent abortions each year in the United States is the solution. Clearly abortionists like Hern aren't so concerned about maternal mortality as they are about keeping abortion legal.

Even if we accept the numbers that the abortion industry uses, pregnancy is still a relatively safe and natural part of human life. Pregnancy isn't like a disease as the abortion industry would suggest. Women's bodies are naturally made to conceive and grow babies. Pregnancy is a function of the female body working normally. Pregnancy does not mean that the female body is abnormal or sick. Even with our maternal mortality crisis, the risk of death from pregnancy is still low. In comparison, pregnancy is only slightly more dangerous than driving a car.[8] And yet most of us drive and ride in cars without a second thought. I'd be highly surprised if Hern or any of his fellow abortionists are avoiding cars because they are so deadly and scary. Cars are a normal part of our lives just as pregnancy is a normal part of many women's lives.

Even if the statistics that the abortion industry uses are correct, it still is not a good justification for nearly a million acts of abortion violence each year. However, I don't accept those numbers. The abortion industry compares two numbers. The first is the maternal mortality rate which is the number of pregnancy-related deaths per 100,000 live births. 17.2 is the latest maternal mortality rate. The second number is the abortion mortality rate, which is the number of abortion related deaths per 100,000 abortions. Hern claims .7 is the abortion mortality rate. But the problem with these numbers is that they are not tracked the same way and are not tracked for the same purposes.

Maternal mortality is tracked differently in a number of ways. Our healthcare system tracks maternal mortality meticulously. This is very important because reducing maternal mortality is a high priority. Our healthcare system needs to track these deaths very closely so that we can lower the rate and save lives.

Deaths that are connected in some way to pregnancy can be tracked up to a year after the birth. So if a woman dies eleven months after the birth and her death can be linked in some way to the pregnancy, then she can be included in the maternal mortality rate. Further, not all maternal deaths result in live birth. But the rate is calculated based on the number of deaths per 100,000 live births. If we took the number of deaths based on total pregnancies, not including those ending in abortion, the maternal mortality rate would be lower.

Abortion deaths, by contrast, are not tracked meticulously. Tracking these deaths is not a priority because reducing the number of abortion-related maternal deaths is not a priority. Most abortionists face no consequences if they fail to report a death. No abortionist wants the publicity from a dead patient. And if a woman goes to the emergency room after an abortion, the abortionist may not even know she died. Further, a woman going to the emergency room may not even reveal that she had an abortion due to not wanting people to know about it. Finally, there are a number of structural problems with our reporting system and with coding that result in abortion-related deaths not being reported as abortion-related deaths. For example, if an abortionist fails to identify an ectopic pregnancy, and the woman dies from an abortion, it is reported as a death due to ectopic pregnancy rather than a death from abortion.[9] There are a number of reasons why an abortion-related death may not be reported and may not be mentioned as the cause of death. All this leads me to the conclusion that we don't actually know what the abortion mortality rate is. And we don't actually know if or how much lower abortion mortality is than maternal mortality.

The final nail in the coffin of pregnancy alarmism is something called the "protective effect of pregnancy" when looking at the total number of premature deaths of women after pregnancy. Studies from Finland, California, and Senegal have found that women who were pregnant are significantly less likely to die prematurely following a pregnancy than women who were not pregnant.[10, 11, 12] In other words, not only is pregnancy not something scary, it may actually be saving women's lives! In Finland, women who weren't pregnant were twice as likely to die prematurely than women who had a live birth in the year following pregnancy. The California study found that "childbirth without any pregnancy losses (abortion or miscarriage) may have a protective effect..." And the Senegal study found that among women aged 20-39, women who had not been pregnant were two to five times more likely to die than those who were pregnant or recently pregnant. These three studies looked at vastly different populations of women but all found that pregnant women were less likely to die.

The Finish and California studies actually looked at the premature death rates of women who had abortions. They both found that women who had abortions were more likely to die prematurely than women who were not pregnant. The Finish study found that post-abortive women were 1.76 times more likely to die prematurely than women who were not pregnant in the year following their pregnancies. They further estimated that post-abortive women were 4.33 times more likely to die from homicide and 3.68 times more likely to die from suicide.[13] The California study found higher rates of premature death among post-abortive women in the eight years following their pregnancies. For example, they found that women who got an abortion for their first pregnancy were 1.62 times more likely to die prematurely than women who gave birth for their first pregnancy.[14]

While these studies tell us that pregnant women are less likely to die prematurely and that post-abortive women are more likely to die prematurely, they only speculate as to the reasons. Why is it that pregnant women are less likely to die? Is it because they are more careful to take care of themselves? Are people more protective of pregnant women? Are post-abortive women more likely to take risks? Clearly more research is needed to answer these questions. Unfortunately, this research doesn't seem to be a priority. What is clear is that pregnancy alarmism is, in fact, alarmism and not based on facts.

Once you get past the propaganda, the crux of the abortion issue becomes very clear. Abortion is a violent act that intentionally kills a little prenatal human. I've shown that very clearly in this book. But if you don't believe that abortion is violence, you don't need to argue with me. You can argue with the abortionists themselves in the next chapter.

Chapter Notes:

1. Paul, Maureen, et al. *Management of Unintended and Abnormal Pregnancy* Hoboken: Wiley-Blackwell, 2009, pp. 358

2. Ibid, pp. 144, 146, 172, 196-200

3. Ibid, pp. 172

4. Ibid, pp. 149-151

5. Hern, Warren. "Pregnancy Kills. Abortion Saves Lives." *The New York Times.* (2019)

    https://www.nytimes.com/2019/05/21/opinion/alabama-law-abortion.html

6. Peterson, E. et al. "Pregnancy-Related Deaths, United States, 2011-2015, and Strategies for Prevention, 13 States, 2013-2017" *Centers for Disease Control and Prevention.* (2019)

    http://dx.doi.org/10.15585/mmwr.mm6818e1

7. Zane, S. et al. "Abortion-Related Mortality in the United States: 1998-2010." *Centers for Disease Control and Prevention.* (2015)

    https://www.ncbi.nlm.nih.gov/pubmed/26241413

8. "Your Chances of Dying Ranked by Sport and Activity" *Teton Gravity Research.* (2019)

    https://www.tetongravity.com/story/news/your-chances-of-dying-ranked-by-sport-and-activity

9. Reardon, David. et. al. "Deaths Associated with Abortion Compared to Childbirth – A Review of New and Old Data and the Medical and Legal Implications." *Journal of Contemporary Health Law & Policy.* (2004) pp. 289

    https://scholarship.law.edu/cgi/viewcontent.cgi?article=1159&context=jchlp

10. Gissler, Mika. et al. "Pregnancy-associated deaths in Finland 1987-1994 - definition problems and benefits of record linkage." *Acta Obstetricia et Gynecologica* Scandinavica (2010) https://doi.org/10.3109/0001634 9709024605

11. Reardon, David. et al. "Deaths associated with pregnancy outcome: a record linkage study of low income women." *Southern Medical Journal* (2002)

    https://www.ncbi.nlm.nih.gov/pubmed/12190217

12. Ronsmans, Carine. et al. "Evidence for a 'healthy pregnant woman effect' in Niakhar, Senegal?" *International Journal of Epidemiology.* (2001)

    https://doi.org/10.1093/ije/30.3.467

13. Gissler, pp. 653

14. Reardon, David. et al. "Deaths associated with pregnancy outcome: a record linkage study of low income women."

AM I KILLING?

Chapter 19

Abortionists Admit that They Are Killing

"Am I killing? Yes, I am. I know that."

Late-Term Abortion Specialist Dr. Curtis Boyd

Late-term abortionist Curtis Boyd said that during a TV interview in 2009. Boyd and his wife own and operate Southwestern Women's Options, one of the few abortion clinics in the country that do abortions throughout all of pregnancy. Boyd later claimed he was "ambushed" by the interviewer. He explained, "They said murder. Murder is a legal definition. I said 'Yes, it's killing, but it is not murder.'"[1]

Clearly, it was important to Curtis Boyd that he made the distinction between the legal definition of murder and the killing that he does. In his opinion, the killing is justified in part because it is legal.

While many in the pro-choice movement have based the right to abortion on the belief that you can't know when life begins, an increasing number of abortion executives and the doctors actually doing the abortions are admitting that they kill living humans. This includes some of the leading abortionists like Warren Hern, Curtis Boyd, Willie Parker, and Leroy Carhart. If you doubt that abortion is killing, you don't have to argue with pro-lifers. You have to argue against the many abortionists who are actually doing the killing and readily admit that they are killing.

Willie Parker is the latest abortion doctor to publically and vocally represent the pro-choice movement. He is an African-American and self-identified Christian doing abortions in the South. He rose to fame in the pro-choice movement after publishing his book, *Life's Work: A Moral Argument for Choice*. His fame, however, was short-lived as he was accused of sexual misconduct in the midst of the Me Too movement. In February of 2019, Dr. Parker debated a pro-life criminologist named Dr. Mike Adams in which Parked admitted many times that he is killing human beings. One particular exchange made the issue very clear:

> Dr. Adams : *How many innocent human beings have you intentionally killed in your life's work?*
>
> Dr. Parker : *I don't know. I don't...*

Dr. Adams : You've lost count.

Dr. Parker : I don't... Ummm.

Dr. Adams : You've lost count.

Dr. Parker : If it's a million...

Dr. Adams : 10,000?

Dr. Parker : 20,000

Dr. Adams : 20,000?

Dr. Parker : What's the difference?

Dr. Adams : What's the difference between 20,000 and 30,000?

Dr. Parker : Yeah. Yeah.

Dr. Adams : 10,000 dead human beings. That's the difference.

Dr. Parker : Ok. Ok. So that's morally more culpable than killing one person?

Dr. Parker goes on to argue that killing human beings is fine so long as he doesn't consider them "persons." But he is happy to concede that he is killing innocent humans.[2]

Another example is that of late-term abortionist Leroy Carhart. Carhart specializes in second and third trimester abortions in his two clinics in Nebraska and Maryland. He is also the Carhart in the famous Supreme Court decision in Gonzales v. Carhart. In this 5-4 decision, the Supreme Court upheld the federal partial birth abortion ban, a procedure that Carhart wanted to continue doing. Carhart recently appeared on a BBC investigative show called *Panorama*. He admits to the interviewer, Hilary Andersson, that the fetus is a baby and that he is killing babies.

Carhart   : *To the fetus it makes no difference whether it's born or not born.*

Andersson : *(stares with a look of shock)*

Carhart   : *The baby has no input in this as far as I'm concerned.*

Andersson : *But it's interesting that you use the word baby because a lot of abortionists won't use that. They'll use the word fetus because they don't want to acknowledge that there's a life.*

Carhart   : *I think that it is a baby. And I tell our... I use it with the patients.*

Andersson : *And you don't have a problem killing a baby?*

Carhart   : *Absolutely not. I have no problem if it's in the mother's uterus.*[3]

In 2000, a Canadian journalist confronted a Canadian abortion clinic owner, Joan Wright, about the pro-life claim that abortion stops a beating heart. Wright responded, "Good grief! They accuse us of pretending we're not doing what we're doing? I'm in the business of death!"[4]

Another more recent example is that of Ann Furedi. Furedi is the CEO of the British Pregnancy Advisory Service, the largest abortion provider in the United Kingdom, performing 70,000 abortions each year. In February of 2019 she debated Sue Thayer, a former Planned Parenthood clinic director turned pro-life advocate. Furedi said:

> *I, too, believe that from the time an egg is fertilized that a human life is there...and that unique life is wonderful and is marvelous...but the point is... it's really about who makes the decision about how that life is valued...in relation to other things.*

Furedi also went on to admit:

> *Women come to us knowing that they are going to have a baby if they don't have an abortion. That is what is growing inside them.*[5]

These are just a sample of many of the abortionists and abortion industry executives who admit that abortion is killing and violent. And the admissions are increasing as more and more people in the abortion industry are embracing a more upfront and unapologetic approach to abortion advocacy.

The pro-choice movement is essentially divided into two camps: those who deny that we can know when life begins and those who admit that they are killing living humans but claim that the killing is justified. The majority of pro-choice people appear to be in the first camp in my experience. I will call this the doubting camp because they doubt that abortion is killing. The doubting camp can be divided into two types of doubters: those who simply don't know when life begins and those who claim that we can't know when life begins. Here again, the majority of pro-choice people appear to fall into the camp of those who simply don't know. This is a somewhat excusable position. The pro-choice movement has seeded much confusion in order to cause people to doubt. Further, we can't reasonably expect every person to have an educated opinion about every issue. If this is you and you are unsure of when life begins, all you must do is look to biologists who are the experts and become more educated on the subject. There is no need to be confused about this question.

Those who claim that you can't know when life begins, however, are not excusable. There is a difference between claiming not to know and asserting that it is impossible to know. If you claim that it is impossible to objectively know, then you should have to defend your assertion. Unfortunately, most people asserting this want to have the best of both camps. They want to be able to assert that we can't know while also claiming ignorance at the same time so that they don't have to defend their position. Those of us who are pro-life should not give these doubters a pass. When they make an assertion, they should be pressed to defend it.

I will call the second camp the anti-egalitarian or anti-equality camp. This is the camp that admits when pressed that abortion is the intentional killing of a living human, but believe that the killing is justified. They generally believe that it is justified to kill these humans because they deny that these humans are equal members of the human family and therefore not deserving of equal rights. For the anti-egalitarian camp, humans don't become equal members of the human family until some later arbitrary point like birth or viability. This camp can typically be identified because they will concede the biology that life begins at conception but will claim that it isn't a "person," isn't a "child," or isn't a "baby." When they make these claims, they aren't using these words to make an argument about the fetus, but to instead deny it equal rights. And so "It's not a baby!" actually means "It is not an equal member of the human family and therefore does not deserve equality." And so the words baby, person, and child simply end up being used as tools or weapons to argue against equal rights.

This anti-egalitarian camp typically and falsely claims that they are taking a scientific position. They only rely on science and biology, however, to identify arbitrary points of development to use as justification for denying equal rights. For example, many use science to describe viability in order to use viability as a reason to deny equal rights to the human before viability. But the anti-egalitarian camp always relies on anti-egalitarian philosophy to justify the killing. They usually try to pretend that their philosophy is part of the science. But philosophy is not science and science is not philosophy. Science can't tell you when killing is justified or not justified. That is a philosophical or religious question. Likewise, philosophy and religion aren't able to tell you when life begins. Only science can do that. And so at the heart of the anti-equality argument for abortion is anti-equality philosophy, often disguised or deceptively presented as science.

Not surprisingly, the anti-egalitarian camp does not want to be labeled as such. Most Americans reject killing and violence as the solution to our problems. And most Americans are vehemently opposed to any philosophy that denies an entire segment of humans their equal rights. And so for pro-life people, all we must do is simply show the truth that abortion is killing and anti-equality. Showing that abortion is inherently violent and anti-equality is 90% of the battle. We can argue against the philosophy that justifies abortion violence. But most Americans already agree with us that violence is wrong.

This is why when I engage a pro-choice person, I usually confront them straight away with the question of violence rather than engaging their pro-choice arguments. Most pro-choice arguments are meaningless if abortion is a violent act. For example, "My body, my choice!" doesn't make any logical sense if abortion is a violent act. Violence is the heart of the issue, but most pro-choice arguments try to steer away from the heart of the issue, not toward it. The goal of the pro-life person should be to go straight to the heart of the issue, that abortion is a violent act, and then to challenge the pro-choice person to explain when and why violence is the right solution. When confronted, most pro-choice people will not dispute that abortion is violence because it is so obviously true. And most pro-choice people will struggle to explain when violence is justified.

For example, I recently engaged a pro-choice Republican at a political event. She bemoaned the fact that abortion was a campaign issue and stated that she thought the government shouldn't fund abortion and should stay out of the issue entirely. I started by agreeing with her on funding by saying that I don't believe the government should fund violent acts. And then I challenged her to tell me if there were any other acts of violence that she thought the government should stay out of. Why should this be the one act of violence that the government should stay out of? I could see her thinking hard on my question, but she ultimately couldn't come up with an answer to my question.

The pro-life position is simple and relatively easy to defend. It is a two part position. The first part is that abortion is a violent act. The second part is that violence is not the way to solve our problems, even in tragic situations. When picking between the side that sees violence as the solution and the side that rejects violence as the solution, I'm going to side with the nonviolent side every single time. And I am confident that most Americans believe in nonviolent solutions as well. I hope that you will join me and the majority of Americans in opposing violence wherever possible. As you will see in the next chapter, abortion violence is fundamentally at odds with America's core values.

Chapter Notes

1. McVeigh, Karen. "I can't think of a time when it was worse: US abortion doctors speak out" *The Guardian.* (2014) https://www.theguardian.com/world/2014/nov/21/us-abortion-doctors-speak-40-years

2. Shepherd, Josh. "WATCH: Abortion Provider Spars With Pro-Life Expert in Explosive University Debate" *The Stream.* (2019)

   https://stream.org/abortion-debate-mike-adams-willie-parker/

   This exchange can be watched at 1 hour and 18 minutes.

3. Bilger, Micaiah. "Late-Term Abortionist Admits: It's a "Baby" and "I Have No Problem" Killing It" *LifeNews.* (2019)

   https://www.lifenews.com/2019/08/01/late-term-abortionist-admits-its-a-baby-and-i-have-no-problem-killing-it/

4. Stern, Leonard. "Abortion Wars" *The Ottawa Citizen* (2000)

5. Carney, Shawn. "Proud of 70,000 abortions a year" 40 *Days For Life.* (2019)

   https://40daysforlife.com/2019/02/19/proud-of-70000-abortions-a-year

DARE TO DREAM

Chapter 20

Why We Should End Abortion Violence

There is a balm in Gilead
To make the wounded whole;
There is a balm in Gilead
To heal the sin-sick soul.

Sometimes I feel discouraged,
And think my work's in vain,
But then the Holy Spirit
Revives my soul again.

If you cannot sing like angels,
If you can't preach like Paul,
You can tell the love of Jesus,
And say He died for all.

Traditional African-American Spiritual
Author Unknown

Ending abortion in America isn't as big of a dream as it first appears. Unfortunately, most people seem to have accepted abortion as just a fact of life. But that's not supported by the data or our history. In fact, legalized elective abortions are only a recent phenomenon in American history. America is about

243 years old as of the writing of this book. But Roe was decided by the Supreme Court only 46 years ago. Further, the abortion rate is at a historic low and continuing to fall. The abortion industry is in crisis as fewer and fewer people choose abortion.

In this chapter we are going to look at recent societal trends and how our culture is moving away from abortion. Then we are going to put abortion in the context of our history of oppression of African-Americans and our nation's core values.

In order to understand how we got here and where we are going, we need to look at the big picture starting with the birth control pill and the sexual revolution. Abortion didn't just come out of nowhere. There were events that led to it. Before the introduction of the birth control pill in 1960, the only methods of birth control were fertility awareness and barriers, especially condoms. The problem with condoms is their high failure rate. About 18 out of every 100 women using condoms will get pregnant in a year. And so casual sex and premarital sex came with a high cost. The risk of getting pregnant and the social and economic costs of pregnancy outside of marriage meant that people were less willing to engage in sex outside of marriage. But then the birth control pill came along with a failure rate half that of condoms. Almost overnight the sexual revolution was born. The birth control pill was just one factor, but it was a very important factor in that people perceived lower risk.

But in reality, the birth control pill had a substantial failure rate. The result was a lot more people having sex with the birth control pill which has a failure rate of 9 out of every 100 women in a year. Not surprising the result was a wave of unwanted pregnancies and along with those unwanted pregnancies a demand for abortion. Women wanted a backup plan in case birth control didn't work. That back up plan was abortion. By the time Roe was decided in 1973, the abortion rate hit 16.3 abortions per 100,000 women. Roughly half of those were illegal and half were legal. A few states had legalized elective abortions before Roe. But it opened the floodgates. In the decade following Roe, the abortion rate nearly doubled. Notice that legalizing abortion caused the abortion rate to go up dramatically, not down. It's really shocking and outrageous that the abortion industry has convinced millions of Americans that legalizing abortion causes the abortion rate to go down. That's simply a lie. The abortion rate did not go down after Roe. It doubled. The abortion rate only started to decline around 1990. The abortion rate continued to rise for 13 years following Roe.

In 1990 that all changed. Not only did the abortion rate begin to fall, it fell dramatically. And it continues to fall to this very day. 2014 marked a very important year since that was the year that the abortion rate fell below that of 1973 when Roe was decided. In other words, for the first time since Roe, the demand for abortion that resulted in Roe just isn't there anymore. This terrifies the abortion industry. Their revenue is at stake. Planned Parenthood has been able to keep their revenue up by taking a greater market share of abortions away from the independent clinics. But more importantly, if increased demand for abortion resulted in Roe, why couldn't a decreased demand result in the overturning of Roe? I think that the decreased demand certainly could lead to the end of Roe. And I think many in the abortion industry would quietly agree with me, even if they don't want to admit it. The reality is that we live in a Democratic Republic. All three branches of our federal government are reactive to the culture. Politicians can pass whatever laws they want now, but the culture always ultimately wins. Politics, entertainment, and media are important parts of the culture. But they are not as important as peoples' everyday decisions about how to live their lives. There is no greater change in the culture regarding abortion than the precipitous and decades-long decline in abortion. It hasn't hit politics yet, but people are voting with their pregnancy decisions. They are voting against abortion.

It is beyond the scope of this book to look comprehensively at the reasons for the abortion decline. And in some ways it really doesn't matter. The fact is that abortion is going out of style. The culture isn't changing back to that of the 1950s, but it isn't staying stuck in 1973 either. It is changing into a new 21st century culture. That culture is clearly rejecting abortion. Among the many factors and proposed factors in this decline include contraceptives, social acceptance of single parenting, the economy and prosperity, pornography, delayed sexual activity in young people, TV shows depicting teen parents, and better ultrasound technology. There is even a hypothesis, inspired by the Donohue – Levitt hypothesis linking abortion to crime, which says that the abortion rate has gone down as a result of what would have been future abortive women being aborted themselves. While I don't know if this hypothesis has much evidence, it would be shocking if the abortion industry was actually aborting away its future customers and putting itself out of business. The pro-choice movement puts a singular emphasis on contraceptives and sex education. They often cite the Affordable Care Act and newly developed hormonal contraceptives as reason for the decline. While I'm sure that contraceptives are a factor, it seems far-fetched to give so much credit to contraceptives when abortion has been in a steady consistent decline since 1990 and our culture has been flush with contraceptives going back to the sexual revolution. The pro-choice movement usually fails to give any consideration to all the other factors while maintaining a singular focus on contraceptives.

There is one glaring factor in the abortion decline, however, that is not tied to contraceptives: the number of young people increasingly delaying sex. Not only is abortion going out of style, but so is sex among young people. Kate Julian wrote a fantastic in-depth article in *The Atlantic* entitled "Why Are Young People Having So Little Sex." She begins, "Despite the easing of taboos and the rise of hookup apps, Americans are in the midst of a sex recession."[1] This is not a conservative or pro-life article. It is published in the progressive magazine *The Atlantic*. If you want to further understand the sexual recession, I'd encourage you to start with this article. There is no doubt that we are in a dramatic sexual recession, one that mirrors the decline in abortion beginning in 1990. Today it is more likely than not that an average high schooler is not having sex. That's really shocking. In fact, the decline in teen sex and in abortion may be one of the greatest recent social advancements in our country, despite the fact that most people are unware of it. In other words, we went from a shocking sexual revolution to an equally shocking sexual recession. But this doesn't fit in to the contraceptive narrative coming from the pro-choice movement and their friends in the media.

The reasons for both of these declines are largely unknown. This falls into an area of social science that nobody seems to be able to pin down. Like many people, I have my own hypothesis which I will share with you. It is a relatively simple idea. I believe that young people are rejecting their parents' life decisions. I think young people are smart enough to see their parents' and their grandparents' dumb mistakes. They can see how much they've suffered from those choices and they simply don't want to make the same mistakes. And so they are going to the other extreme. Not only are they putting off sex and rejecting abortion, they are also putting off marriage or rejecting marriage entirely. Am I giving young people too much credit? Maybe I am, but I don't believe so. Are young people overreacting? Maybe in some ways they are. I love being married. I wish everyone could have a marriage and parenting experience that is as fulfilling as mine. But it seems that many young people have checked out of that whole idea.

Regardless of how you see it, the future is very bright for the pro-life movement. We have never been in a better position to make gains across all areas. We will win because the culture is changing. It's not changing back to the predominantly Christian culture of the 1950s or the abortion culture of the 1970s. It is a new 21st century culture. It is a different culture. It is unbelievably prosperous. It is more supportive of single parents than ever before.

Ultimately, I believe we will win because Americans believe in equality and reject violence. What you will see in the last section of this book is that America was founded on certain core values and that these values cannot be consistent with abortion violence.

On Saturday September 26, 2015, Felicia and I took a day trip to relax on Jekyll Island just off the coast of Brunswick, Georgia. On Monday, the baby we were adopting was to be delivered via a scheduled cesarean section. But on that Saturday we were trying to calm our nerves and mentally prepare. Jekyll Island is a popular vacation destination today. But it also has a long history of slavery and racial segregation. And at one time it was the location of some of the last gasps of the trans-Atlantic slave trade.

In the mid-19th century, Jekyll Island was owned by the DuBignon family. They were slaveholders and proponents of slavery. They also conspired to illegally smuggle slaves from Africa. This was the time period right before the American Civil War. Slavery was still legal in the southern states, including Jekyll Island. The trans-Atlantic slave trade, however, was illegal in America, including the south. This didn't stop proponents of slavery from trying. They aggressively believed it was their right to purchase slaves in Africa and bring them to America. One such proponent was Charles Lamar. Lamar conspired with the DuBignons and a man named William Corrie, the owner of a pleasure yacht named *Wanderer*. Together this group of slavery proponents plotted and successfully carried out a plan to buy slaves at the mouth of the Congo River. The ship was retrofitted with a secret deck for the slaves and a secret fresh water tank. But they needed a private place to unload the slaves away from prying eyes. Jekyll Island provided just the place. *Wanderer* was the last illegal slave ship to deliver to Georgia and one of the last slave ships to deliver to America. The federal government tried and failed to prosecute the group. Not long after, the nation would enter into the most deadly war in American history. Ironically, the *Wanderer*, which was used to enslave over 400 people, was converted by the Union Army into a war ship to free those same slaves.[2]

For my family, this story is now part of our story. On the Monday after our visit to Jekyll island, our son Derrick was born just a few miles away. Unfortunately, we don't know his family tree. We don't know the true story of his ancestors and how they came here. But what we do know is that nearly all of his DNA comes from that region of Africa. We also know that in the county in which Derrick was born, 72% of the population were slaves in 1860, immediately before the civil war.[3] And so the history of slavery and segregation has become a part of my family's history through adoption. But America's story doesn't end there. In the span of about 240 years, we went from being a country of slavery to a country of segregation and Jim Crow oppression to a country where my wife and I were able to adopt a child of another race and be accepted as a mixed race family. America was transformed into a country where we are accepted and supported by our church, neighbors, and community.

During the period before the civil war, American abolitionists could be divided roughly into two groups. They were the followers of William Lloyd Garrison who believed that America was founded on slavery and those like Frederick Douglas who believed later in his life that America was founded on principles that were inherently inconsistent with slavery. For Garrison, our Constitution was a pro-slavery document, America was founded as an evil country, and the American government was beyond redemption. They swore off civic activities such as voting and politics. But the former slave and popular public speaker Frederick Douglas came to disagree with Garrison. Instead, he believed that America's founding was good and right. He defended the core values that America was built upon. These core values were best summed up by the founding father George Mason in the Virginia Declaration of Rights which inspired the Declaration of Independence. He wrote, "That all men are by nature equally free and independent and have certain inherent rights, of which, when they enter into a state of society, they cannot, by any compact, deprive or divest their posterity; namely, the enjoyment of life and liberty, with the means of acquiring and possessing property, and pursuing and obtaining happiness and safety."[4] And so we can see that included in our core values are life, liberty, and equality.

Douglass used his soaring rhetoric to call Americans back to our core values which were at odds with slavery. In essence, he made out slavery to be Un-American. This can be best seen in a famous Independence Day speech he gave in 1852 that is often given the title "What to the Slave is the 4th of July?" I was struck as I read his speech how he went out of his way to celebrate the founding fathers, the Declaration of Independence, and the Revolutionary War. For example, he said, "The signers of the Declaration of

Independence were brave men. They were great men too – great enough to give fame to a great age. It does not often happen to a nation to raise, at one time, such a number of truly great men." Douglas also said, "They were statesmen, patriots and heroes, and for the good they did, and the principles they contended for, I will unite with you to honor their memory." And near the end of the speech he drove the point home saying, "I, therefore, leave off where I began, with hope. While drawing encouragement from the Declaration of Independence, the great principles it contains, and the genius of American Institutions..."[5]

But Douglass didn't just praise the core values that birthed our nation. He ruthlessly attacked slavery as a violation of our core values saying:

> What, to the American slave, is your 4th of July? I answer: a day that reveals to him, more than all other days in the year, the gross injustice and cruelty to which he is the constant victim. To him, your celebration is a sham; your boasted liberty, an unholy license; your national greatness, swelling vanity; your sounds of rejoicing are empty and heartless; your denunciations of tyrants, brass fronted impudence; your shouts of liberty and equality, hollow mockery; your prayers and hymns, your

*sermons and thanksgivings, with all your
religious parade, and solemnity, are, to him,
mere bombast, fraud, deception, impiety, and
hypocrisy — a thin veil to cover up crimes
which would disgrace a nation of savages.
There is not a nation on the earth guilty of
practices, more shocking and bloody, than are
the people of these United States, at this very
hour.*[6]

What Douglass did, calling Americans back to our core values, has become an inspiration and strategy for Americans fighting violent injustice to this very day. Douglass may not have been the first human rights activist to use this strategy, but he used it so effectively that we remember him today. One of the reasons it was so powerful is because America is a nation whose identity is rooted in our core values from our founding. Our identity is not tied to ethnicity. This isn't the case for many countries around the world whose identities are in some way tied to ethnicity. Russia is of the Russians. Japan is of the Japanese. Ireland is of the Irish. But America is different. America is a nation of liberty lovers. America is about being a place where people can get a fresh start and build their own future. America is about self-determination and individual rights. But most importantly, America is about giving everyone a fair shake, equality and justice for all! America isn't special because we got it right. Clearly we got it wrong over and over again. And we continue to get it wrong with the widespread and bloody violence of abortion. But America is special because we aspire to get it right from our very founding. So many times we have gone back to our core values to end oppression. We must keep going back to our core values if we are to continue to be Americans, liberty loving people.

111 years after Frederick Douglas gave his "What to the Slave is the 4th of July?" speech, another African-American man, Dr. Rev. Martin Luther King, stood on the steps of the Lincoln Memorial and gave another speech that shook the nation. King's "I Have a Dream" speech also appealed to America's core values in order to end segregation and the violence of the Jim Crow south. His inspiration came from a phrase that was popularized just a few decades earlier, the "American Dream." For so many Americans, the American Dream became a phrase that summed up all of our core values. King understood that and he wanted all Americans to understand that African-Americans were not being included in their dream.

King used the analogy of a bad check to describe America's core values which were being withheld from African-Americans. He preached:

In a sense we've come to our nation's capital to cash a check. When the architects of our Republic wrote the magnificent words of the Constitution and the Declaration of Independence, they were signing a promissory note to which every American was to fall heir. This note was a promise that all men – yes, black men as well as white men – would be guaranteed the unalienable rights of life liberty and the pursuit of happiness. It is obvious today that America has defaulted on this promissory note insofar as her citizens of color are concerned. Instead of honoring this sacred obligation, America has given the Negro people a bad check, a check which has come back marked "insufficient funds."[7]

In King's powerful analogue, America was founded on a promise to defend the natural rights of all its citizens. This promise was represented by a check that was bounced. King had the courage to accuse America of not keeping its word, of bouncing a check. There are very few things that are as valuable to a man as his word. To attack a man's word is to attack his honor and respect. Essentially, King attacked America's honor. He embarrassed us where we so greatly deserved to be embarrassed. And it worked. Real men keep their promises. And a genuine America keeps its promise to defend the natural rights that our country was nobly founded upon.

King went on to give his people desperately needed hope in the now infamous "I have a dream" portion of his speech. He started with "It is a dream rooted deeply in the American dream. I have a dream that one day this nation will rise up, live out the true meaning of its creed: We hold these truths to be self-evident, that all men are created equal." King went on to describe five pictures of his dream. He was painting a picture of what America could be. He was painting a picture of a liberty worth pursuing even when things seemed hopeless. The picture he painted was powerful, but in its time seemed far-fetched. Could the children of slaves and slave owners really eat a meal together? Would Mississippi really be transformed into an oasis of freedom? Would our nation really judge King's children by the content of their character? King ends his dream with a picture of Alabama, the very epicenter of police brutality and bombings, being a place where black boys and girls and white boys and girls hold hands as brothers and sisters in a mutual human family.[8] Could this really be possible? Or was King just being a silly dreamer?

But what I discovered was that not only was King not a silly dreamer, in fact he didn't dream enough!

King described black and white children as part of a human family. But even King didn't dare to dream of white children and black children actually being brothers and sisters. And yet my son is black and my girls are white. And our family isn't even unusual. I know many families just like mine where black and white kids are truly siblings. In his time, King's dream may have seemed far-fetched. If only he could have lived to see my family! While we aren't finished yet in the cause of racial justice, we have transformed further in 50 years than even the Reverend King could have dared to dream.

And yet with all the reasons to celebrate the cause of equality and justice for all, there is even more reason to grieve and be embarrassed. We didn't simply end segregation. Soon after the civil rights movement, we embraced a new violent oppression: abortion. How can we boast equality, after replacing the sweltering heat of segregation with the chilling cold steel of the curette blade? How can we claim inalienable rights, after trading the nooses used to hang black bodies with the vacuum used to dismember tiny fetal bodies? How can we claim to be the people of liberty, after trading the biting teeth of Bull Connor's attack dogs with the sharp steel teeth of the abortionist's forceps? How can we hold up the banner of "life, liberty, and the pursuit of happiness" after replacing the bombs and the burnt bodies of four little girls at the 16th Street Baptist Church with the countless little fetal bodies fatally burned by poisonous saline injection? How dare we utter the word "freedom" while unfathomable numbers of severed arms, legs, heads, and torsos of prenatal children have been incinerated, ground up, and flushed in the name of liberty! We are certainly the most pitiful of nations. We dare to call ourselves "the land of the free" and we enjoy the most lavish prosperity on earth, but then we engage in such bloody and grotesque violence. No nation on earth has less of an excuse than us. We of all people know better!

Despite our hypocrisy, we still believe in inalienable human rights, including life, liberty, and the pursuit of happiness. We don't believe in born rights as if birth gives you rights. We don't believe in viability rights as if some subjective label of viability gives you rights. We believe in human rights which you possess simply by being human. I am daring to dream and am inspired by the dreams of Frederick Douglass and Martin Luther King. I dream of a day when abortion isn't only illegal, but when no one would even consider it an option. I dream of a day when no woman has to choose between her child and her job or school. I dream of a day when we make it easier, not harder, to be a single parent. I dare to dream of the day when every single human, regardless of circumstance, is afforded equal and inalienable human rights and is treated with dignity and respect. And of course I dream of a day when there is a universal consensus that abortion is a violent and unacceptable act. Am I a silly dreamer? Maybe I am, but I dream anyway, just as the Rev. King taught us. I invite you and even dare you to dream with me. If you believe in our core values, this dream is for you. This is our American dream.

Chapter Notes:

1. Julian, Kate. "Why Are Young People Having So Little Sex?" *The Atlantic*. (2018)

   https://www.theatlantic.com/magazine/archive/2018/12/the-sex-recession/573949/

2. Jones, Tyler. "On Jekyll Island, black history remains prominent" *The Brunswick News*. (2017)

   https://thebrunswicknews.com/life/on-jekyll-island-black-history-remains-prominent/article_3272b285-ee8e-5f5a-a7c7-a0683a9bbb6f.html

3. Hergesheimer, E. "Map Showing the Distribution of the Slave Population of the Southern States of the United States Compiled from the Census of 1860"

   https://upload.wikimedia.org/wikipedia/commons/5/5e/SlavePopulationUS1860.jpg

4. Mason, George. "The Virginia Declaration of Rights" *Virginia Constitutional Convention*. (1776)

   https://www.archives.gov/founding-docs/virginia-declaration-of-rights

5. Douglass, Frederick. "Oration, Delivered in Corinthian Hall, Rochester, By Frederick Douglass, JULY 5TH, 1852."

   https://rbscp.lib.rochester.edu/2945

6. Ibid

7. King, Martin Luther. "I Have A Dream..." Speech by the Rev. Martin Luther King at the "March on Washington" (1963)

   https://www.archives.gov/files/press/exhibits/dream-speech.pdf

8. Ibid

NOTE FROM THE AUTHOR

I want to thank you for taking the time to read *Fetal Beauty*. This is a difficult subject to write and read about. It is often unpleasant and even disturbing. I appreciate that you took the time to read about this subject and that you choose *Fetal Beauty*. Regardless of your position on abortion, I hope you've come away from this book better informed and better prepared to make decisions in your own life. If you think that *Fetal Beauty* is a worthwhile read, would you do a favor for me? Would you recommend *Fetal Beauty* to your friends and family? I would also greatly appreciate a positive review on Amazon. We each can do our part in making a less violent society.

Sincerely,

Jordan Warfel